MANCHESTER GERMAN TEXTS

Pioniere in Ingolstadt

MANCHESTER GERMAN TEXTS

also available

Braun (Hollis ed.) *Unvollendete Geschichte*
de Bruyn (Tate ed.) *Märkische Forschungen*
Peitsch, Williams (eds.) *Berlin seit dem Kriegsende*
Wallraff (Sandford ed.) *Der Aufmacher*
Böll (Reid ed.) *Was soll aus dem Jungen bloß werden?*

forthcoming titles include:

Pausewang (Tebbutt ed.) *Die Wolke*

of related interest:

Williams, Peitsch (eds.) *German literature and the Cold War**
Büchner (Jacobs ed.) *Dantons Tod und Woyzeck*
Townson *Mother-tongue and fatherland*
Hall *Modern German pronunciation*

*hardback only

Marieluise Fleißer

Pioniere in Ingolstadt

edited with introduction, notes and vocabulary by

David Horton

Universität des Saarlandes, Saarbrücken

Manchester University Press

Manchester and New York

distributed exclusively in the USA and Canada by St. Martin's Press

Published by Manchester University Press
Oxford Road, Manchester M13 9PL, UK
and Room 400, 175 Fifth Avenue, New York, NY 10010, USA

Distributed exclusively in the USA and Canada
by St. Martin's Press, Inc., 175 Fifth Avenue, New York, NY 10010, USA

British Library cataloguing-in-publication data
A catalogue record for this book is available from the British Library

Library of Congress cataloging-in-publication data
Fleisser, Marieluise, 1901–1974
Pioniere in Ingolstadt / Marieluise Fleisser; edited with an introduction and notes by David Horton
p. cm. – (Manchester new German texts)
Includes bibliographical references.
ISBN 0-7190-3467-1 (cloth). – ISBN 0-7190-3468-X (paper)
I. Horton, David. II. Title. III. Series.
PT2611.L46P5 1992
832'.914–dc20 91-31668

ISBN 0 7190 3467-1 *hardback*
0 7190 3468-X *paperback*

Typeset in Times
by Koinonia Ltd, Manchester
Printed in Great Britain
by Bell & Bain Limited, Glasgow

Contents

Series preface

The *Manchester German texts* series has been devised in response to recent curricular reforms at school and undergraduate level. A major stimulus to the thinking of the editorial board has been the introduction of the new A level syllabuses. The Manchester editions have accordingly been designed for use in both literature and topic-based work, with the editorial apparatus encouraging exploration of the texts through the medium of German. In addition to the features normally included in an advanced Modern Languages series, the editions contain a new and distinctive section entirely in German called the *Arbeitsteil*. It is envisaged that the Manchester editorial approach, in conjunction with a careful choice of texts and material, will equip students to meet the new demands and challenges in German studies.

Acknowledgements

I am indebted to the following for their kind support in the preparation of this edition, in particular for their invaluable assistance with the clarification of a number of queries: Elisabeth Bond-Pablé, co-translator of the play, with Tinch Minter, for the production at the Gate Theatre, London in 1991; Klaus Gültig, Marieluise Fleißer's nephew; and Frau Eiden of the Wissenschaftliche Stadtbibliothek, Ingolstadt, home of the Marieluise-Fleißer-Archiv. Not least I am extremely grateful to my wife, Mina, for patiently answering a great many questions on matters of German stylistic usage.

Introduction

The place of Marieluise Fleißer (1901-74) in the history of German literature has been an unstable one. A child of the 1920s, she rose to fame as a protégée of Bertolt Brecht, who arranged the scandalous second production of her play *Pioniere in Ingolstadt* in Berlin in 1929, and on the strength of a collection of stories published in the same year. But her promising literary career was cut short by the rise of the National Socialists, who burned her books, by the upheaval of the Second World War, and by a disastrous marriage. By the 1940s she had sunk into virtual oblivion. It was not until the 1970s that Marieluise Fleißer came to occupy the place she now holds as a major German writer of her generation. The works which had attracted attention in the 1920s, especially *Pioniere in Ingolstadt*, were rediscovered by a new generation of German playwrights, who reassessed her contribution to social drama and paved the way for a Fleißer renaissance which has continued to this day. It was in the 1970s that her plays and stories gained widespread acclaim, and that the collected edition of her works appeared. Although her works do not conceal their historical roots in the world of the Weimar Republic, they are surprisingly contemporary, as their fortunes over the past twenty years confirm.

Recently translated into English, *Pioniere in Ingolstadt* received its British première at the Gate Theatre, London in February 1991, where it was greeted with considerable critical acclaim.

Marieluise Fleißer: life, work, reception

The life of Marieluise Fleißer: 'in die Enge geht alles'[1]

Marieluise Fleißer was born in the Bavarian town of Ingolstadt on 23 November 1901, and died there on 2 February 1974. At her birth, Ingolstadt was a provincial backwater of some 15,000 inhabitants, dominated by the spires of its nine churches, still medieval in outline, and economically dependent on its large garrison. By the time of her death it had become a thriving town of some 85,000 people, a cultural entity in its own right, and an important centre of the oil and automobile industries. It was here that Fleißer spent over fifty years of her life, and it is with Ingolstadt that her name has always been linked, providing not only the titles of her most famous works, but frequently also their settings, their streets and churches.[2]

Fleißer's relationship with her home-town was, however, by no means as unambiguous as the length of time she spent there might suggest. Indeed, she spent some fifteen years as a young woman attempting to liberate herself from the restrictions of the provincial town, with its narrow-minded conservatism, philistinism and oppressive Catholicism. After schooling in Ingolstadt and, from 1914 to 1919, at a Catholic boarding school in Regensburg, she sought out the cosmopolitanism of larger cities, going on to study mainly drama at the University of Munich (1919-25), and subsequently attempting to establish herself as a writer in the metropolis of Berlin. In the city she tried to cast off the stifling limitations which a provincial childhood and above all the (in her view) unnatural conditions of a convent education had imposed upon her. 'Das Kloster war für mich eine Kaserne oder ein Gefängnis', she wrote in 1973 (IV, 496), and she continued to dwell throughout her life, in works, interviews and letters, on the complexes and distorted notions which her Catholic 'Scheuklappenerziehung' (III, 312) had instilled indelibly within her, an intelligent and critical young woman.[3]

The beginnings of Fleißer's attempts at liberation were promising. Once in Munich, she soon left the convent accommodation her father had found for her, settled in the artists' quarter of Schwabing, and met her first lover, Alexander Weicker ('Jappes'), a third-rate novelist of questionable character. These were exciting, if materially difficult, years. She began to write, and gained access to avant-garde circles, moving in the company of talented and radical intellectuals. In 1924 she made the acquaintance of the young poet–playwright who was to prove the most powerful influence on her life, Bertolt Brecht. As she later conceded in a number of writings, most notably in the story *Avantgarde* (written 1962/63), Fleißer was fascinated, indeed utterly overwhelmed, by the arrogant and rebellious genius of the man who was already on his way to becoming Germany's most significant dramatist: 'Der Mann war eine Potenz, er brach sie sofort' (III, 117). Brecht furthered her career, finding her a publisher and arranging the première of her first play *Fegefeuer in Ingolstadt* in Berlin in 1926. He gave her the idea for *Pioniere in Ingolstadt*, in whose progress he took a particular interest. She become not only his protégée, but also his lover, and spent much time with him between 1924 and 1927, until she finally terminated their relationship following the scandal surrounding the production of *Pioniere in Ingolstadt* in Berlin in 1929. By this stage she had become disillusioned with the insincerity and ruthlessness of the urban intelligentsia, especially of Brecht himself, and returned to Ingolstadt, where she became briefly engaged to the tobacco merchant and sportsman Josef ('Bepp') Haindl. Before finally capitulating to Haindl's intense pressure to marry (which she did in 1935), she made one further attempt to find a place among artistic circles in Berlin, this time through her relationship with the eccentric nationalist journalist and poet

Hellmuth Draws-Tychsen, to whom she was engaged between 1929 and 1934. When this relationship, too, collapsed, she finally returned to Ingolstadt.

For the rest of her life, Fleißer was torn between the two poles of intellectual emancipation and the need for domestic security, between city and town. Desperate to free herself from the frustrating provincialism of life in Ingolstadt, she had sought emancipation in Munich and Berlin only to become confused, disorientated, and even frightened by her new-found freedom ('Etwas Mörderisches war in dieser Freiheit', III, 147). She had learned that, for all its apparent liberalism and egalitarianism, the progressive city society of the 1920s offered little real freedom for the independent female who wished to live by her writing. She had suffered isolation, humiliation and despair, even being brought to the point where she attempted suicide in 1932. Her experiment in self-liberation having failed, she fled back to the rather questionable security of life in Ingolstadt, destined to remain as much an outsider there as she had been in the city.

When she did return to Ingolstadt in 1933, it was in disgrace. She was penniless and without qualifications (she had not completed the studies for which she had left the town some fourteen years earlier). As a result of the *Pioniere* scandal she was shunned by the local people, and her physical safety was threatened by the Nazis, who had come to power in the same year. For a time her father refused to allow her into the parental home. It was in this ignominious situation that, against her better judgement but in dire need of stability, she married Bepp Haindl in 1935.[4] In so doing she condemned herself to a life of bourgeois mediocrity, intellectual renunciation, and physical hardship: forced by her husband to work in the family business, and later dragged into the war effort as a 'Hilfsarbeiterin', she found it almost impossible to continue writing. Furthermore, her earlier books were burned by the Nazis, who also imposed a partial prohibition on her further literary activity. In 1938 she suffered a nervous breakdown.

After the war she continued to write intermittently, and, despite adverse circumstances, managed to produce a small number of valuable works, including the 'Volksstück' *Der starke Stamm*, which proved a moderate success in 1950. But these were sad and difficult years, and between 1949 and 1962 she published nothing.[5] Briefly the idea of escape surfaced once more in her mind, again via Brecht. The two had met again in 1950, when Brecht had instigated the première of *Der starke Stamm* in Munich, and in 1956 he invited her to settle and work in East Berlin. Loyalty to her now sick husband and a (still rather uneasy) attachment to her home-town led her to decline his offer. She remained in virtual oblivion in Ingolstadt, surviving the death of her husband and a heart attack in 1958, until her fortunes took a positive turn with the rediscovery of her earlier works by a new generation of playwrights in the late 1960s.

Unexpectedly, in the wake of the neo-realistic popular drama which now emerged in Germany, she again found herself at the centre of public attention. She set about revising her early works, her plays were frequently performed, and her stories reprinted. Thus it was that by the time she died in 1974 Marieluise Fleißer had once again become, after a trying and emotionally turbulent life, Ingolstadt's most famous daughter.

The works of Marieluise Fleißer: 'immer nur etwas zwischen Männern und Frauen' (Mat., 349)

The works of Marieluise Fleißer are characterised by a striking restriction in scope and a recurrent preoccupation with a small number of themes. Indeed, so limited was the field of Fleißer's imaginative vision that she not only returned almost obsessively to particular ideas and episodes in various works, but also produced numerous different versions of some of her texts, rearranging, deleting or adding passages, and changing their titles. The thematic and conceptual restriction of Fleißer's work is due, however, not so much to a lack of creative power on her part as to a willed concentration on issues which could be explored through a narrow but intense focus. 'Ich könnte natürlich immer nur etwas zwischen Männern und Frauen machen', she said in an interview in 1971, and in so doing identified the one fundamental concern which underlies, in a rich variety of forms, virtually all of her work: the world of private human relationships. Although her life spanned the monumental events of twentieth-century German history – two world wars; the inflation, depression and unemployment of the ill-fated Weimar Republic; the rise of National Socialism and the Third Reich; the destruction and division of the nation; the emergence of a democratic West Germany as an economic and political force – it is the relationship between the sexes, always experienced as a struggle and destined to end (at least for the woman) in bitter disillusion, which provides Fleißer with an inexhaustible fund of material. As was noted above, her own relations with men were highly traumatic,[6] marked by a protracted, and ultimately abortive, search for freedom within security, for self-realisation, both emotional and intellectual, within the context of mutual understanding, support and trust. Her hope, far less self-evident in the 1920s than it might be today, was for a partnership on equal terms. The result of her search was a constant struggle against indifference, even brutality, against exploitation and sheer prejudice.

It is precisely this female quest for self-fulfilment through love and acceptance – in many ways a curiously unemancipated ideal for one as independent as Fleißer – that forms the backbone of her fictional world, especially in the 1920s. The works produced at this time, the most creative period of her career, are those on which her fame principally rests and to which she repeatedly returned. They are autobiographical, even confessional in nature, and constitute a moving testimony to the writer's own attempts to find personal fulfilment. Virtually all of them can be traced back to men and events of her own experience: her relationships with Jappes, Brecht, Haindl, Draws (and others) recur time and again as their thinly disguised background. As the recent volume of her previously unpublished work shows, she was producing stories, film-scripts and radio plays based on episodes dating back as far as 1919 even during the late 1960s, the final creative phase of her career. The entire *oeuvre* of Marieluise Fleißer, then, might be read as a persistent exploration and re-exploration of the circumstances of her own life.

The common thread which runs through Fleißer's early work, of which *Pioniere* is so central a part, is the sensitive exploration of the intensely private sphere of the vulnerable young woman exposed to the harshness of a male-dominated world. It is, though with significant modifications, the story of Fleißer's own adolescence.[7] The girls in these stories are inexperienced, naive and insecure, ill-prepared by their family background, and in some cases by their Catholic upbringing, for the trials of adult life: 'Ich wußte nicht, wie man klug ist. Ich wußte bloß, daß ich aufgewachsen bin in einem Kloster und daß alles, was ich dort gelernt habe, für mein Leben falsch ist' (III, 33). Often alone in unfamiliar surroundings, suffering material deprivation due to the depression (inflation, unemployment), hardly able to feed themselves, let alone dress as they would wish, the protagonists of these stories crave the emotional fulfilment and social integration which only a relationship with a man can provide: 'sie war die unselbständige Person und mußte immer wen hinter sich haben' (III, 66). They seek love and warmth (variously described as 'Hilfe', 'Halt', 'Stütze') as a protection against a harsh world in which humanity and solidarity have been ousted by rivalry and aggression. In their desperate desire to please, these girls are prepared to subordinate themselves utterly to the interests of the dominant male, to show a selfless devotion which frequently takes the form of complete self-denial and degradation. They become martyrs to their own silent suffering, accepting the necessity of pain as a duty, and casting the male in the role of a quasi-divine redeemer who might bring 'Gnade'.[8] The men in these stories – generally unfeeling, egocentric and unashamedly patriarchal in their attitudes – exploit their position, both materially and sexually. They remain calculating and insensitive to the girls' deeper needs, and react aggressively to female dependence (on which, ironically, they depend), frequently abandoning their partners when they have no further use for them. The girls see their hopes shattered, are plunged into despair, and taught the cruel lessons of a world in which human relations are truly a battle: 'Man nannte erwachsen, wem ein Licht aufgegangen war über die natürliche Feindschaft unter den Menschen' (III, 24). In the struggle for domination (life is a 'Schlachtfeld', III, 81), weakness, especially that of the woman, is ruthlessly exposed and exploited: the laws of the social jungle prevail. These females are resigned to their fate, and in one case the girl takes her own life (*Ein Pfund Orangen*). Only in a few of Fleißer's works – they are the later ones – do we meet women who are able to fight back (the novel *Eine Zierde für den Verein*, the plays *Der Tiefseefisch* and *Der Starke Stamm*, anticipated to an extent in the story *Die Ziege*). For all these women, fulfilment is a myth. Fleißer's early stories have aptly been called 'parables of quiet despair'[9], 'Untergangsmärchen aus dem Alltag'.[10]

Whilst the works of the 1920s are autobiographical in the sense that they are loosely based on authentic experience, Fleißer's later works project the theme of 'etwas zwischen Männern und Frauen' very clearly against the back-cloth of her experiences with Brecht, Haindl, Draws and, to a lesser extent, with the teacher Georg Hetzelein in the mid-1930s. Here the concern with female self-realisation and emancipation, central to Fleißer's life at the time, though only implicit in the early stories (with the exception of *Die Ziege*), is fully developed. Taken together, these works (the novel, the *Tiefseefisch* play, the stories of the 1940s and 1960s) read like a commentary on Fleißer's personal attempt to find, in Munich and Berlin, Brecht and Draws, the scope and freedom denied her by Ingolstadt and Haindl. They also read, though, like a paradigm of the sexual politics of their age, a reflection of the difficulties encountered by the educated and independent female in her struggle against social prejudices and gender stereotypes at a time when women had theoretically gained equality (they were given the vote in 1918), but in practice remained subordinate.[11] As was pointed out above, Fleißer remained everywhere an outsider. As a result of convictions instilled in her by her lower middle-class upbringing she was repelled by the moral indifference of the avant-garde, and was shocked to discover that the place of the woman in such circles was only superficially different from that in the patriarchal society from which she had sought refuge. Yet, having tasted the excitement of relative freedom, she was never able to readjust to provincial conditions, as the remainder of her life demonstrated. She felt unable to make a complete break with traditional roles and was caught uncomfortably between the two stools of conformism and individualism. It is significant that she paid homage in a number of works to a handful of truly exceptional figures, all of them men, who did achieve complete emancipation from 'normality', accepting the dangerous consequences: the writer Jean Genet, the artist Van Gogh, and the mountaineer Hermann Buhl. It was this radical rejection of restrictions and norms which she found so fascinating in Brecht: he represented everything she could not become. Marieluise Fleißer's own life, and work, is a tale of frustrated individualism, a feminism *ex negativo*. Equally significantly, she re-embraced Catholicism later in life, seeking here, too, the stability which she so desperately needed ('mein Leben war so, daß ich es nicht mehr ausgehalten habe ohne einen Herrgott').[12]

What are the reasons for Fleißer's persistent, and apparently rather narcis-sistic, preoccupation with her own life, and, within that, for her exclusive concentration on male–female relationships? The answer to this question takes us to the very centre of her (rather vaguely formulated) aesthetic. She was a writer who depended heavily on concrete, lived experience for her art, unable to produce literature from pure imagination or empty abstraction. Repeatedly in articles and interviews she stressed the authentic basis of her creative work as an observation on tangible human reality, 'eine wirkliche Lebensbeobachtung', rejecting the 'Gedankenspiele' and 'Künstelei' of modern art in favour of 'die echten Berührungspunkte' (IV, 541): 'Aber man schreibt doch immer aus dem heraus, was man selbst erfahren oder aus lebendiger Nähe beobachtet hat, man muß es irgendeinmal gekriegt haben, von nichts kommt nichts, jedenfalls bei mir' (*Mat.*, 343). It is striking how often in her isolated comments on her own works she stresses the notion of observation, how she places human interest, character, above all else: 'für mich ist das Ausschlaggebende beim Schreiben der Mensch' (*Mat.*, 349). Aware that this approach to literature had become somewhat unfashionable during her own lifetime, she nevertheless insisted on authenticity of experience as the foundation of her art. She strove to create 'Menschen mit einem ausgesprochenen Eigenleben' (Mat., *357*), writing with great personal involvement.[13] Hence the intense, personal quality of her work, in which she seeks to enter sympathetically into the minds of her characters while at the same time retaining a degree of ironic detachment from them.[14]

The private and autobiographical quality of Fleißer's work does not, however, mean that it is lacking in more general relevance. Her highly personal stories are clearly embedded in the socioeconomic context of their time, which, though rarely explicitly articulated or analysed, conditions the lives and attitudes of her characters. Prevailing social structures and ideologies are everywhere apparent beneath the surface of her plots, setting the parameters within which her characters behave and respond, and which, at times, they seek to transcend. The oppressive conservatism of provincial attitudes and the claustrophobia of rigidly patriarchal structures; the rivalries and aggressions of young people in their (unconscious) process of socialisation; the strict hierarchies of social power; the collapse of the family unit as an anchor of social stability – all these are in evidence as the background to her texts. So, too, is the depressed economic situation of the early 1920s, with inflation, unemployment, outright poverty (*Der Apfel, Ein Pfund Orangen, Moritat vom Institutsfräulein*). Her individual 'parables of quiet despair' provide, then, through the demonstration of the females' problems and their element of critical detachment, a comment, though a veiled one, on society as a whole: 'Unwillkürlich von der Gesellschaft schreiben, eben weil man mit dem Röntgenblick von sich selber schreibt'.[15] Nowhere does she directly voice her dissatisfaction with the state of society, show open rebellion, or point to utopian alternatives. Her protest is more subtle and more restrained, it is implicit rather than overt, though it is no less powerful for that.

The reception of Marieluise Fleißer: 'Und doch wurde ich eine wirkende Kraft' (IV, 503)

In terms of the reception of her work, Marieluise Fleißer might be said to have enjoyed two careers, separated by some fifty years of critical neglect. Both phases of her career were based on the same texts. In the 1920s she became a respected, and subsequently notorious, literary figure. Her first play, *Fegefeuer in Ingolstadt*, was performed to some local acclaim in Berlin in 1926, while her second, *Pioniere in Ingolstadt*, achieved a moderate success upon its first performance in Dresden in 1928. Brecht had a hand in bringing about both productions. In 1929 her first collection of stories, *Ein Pfund Orangen*, was well received, as was her novel, *Eine Zierde für den Verein*, in 1931. But it was the second production of *Pioniere* in Berlin in 1929, for which Brecht made drastic and provocative alterations to the text, ultimately against Fleißer's own wishes, that established her as a national figure. The sexual frankness, indeed calculated obscenity, of Brecht's production caused a *succès de scandale*, one of the major theatrical sensations of the Weimar Republic. Once the furore had subsided, Fleißer, unable to write for the reasons outlined above, sank into oblivion.

Ironically, while it was Brecht who had promoted Fleißer's career in the 1920s, it was a group of playwrights impatient with Brechtian models who brought about the Fleißer renaissance of the 1970s. Disturbed by what they perceived as a contrived utopianism and an excessively high degree of analytical reflection in Brecht's plays, a group of young Bavarian writers began to explore an alternative tradition of social drama, a tradition in which social issues are articulated through the restricted language and consciousness of uneducated, provincial people. Ödön von Horváth (1901-38) and Marieluise Fleißer were the main beneficiaries of this development, which produced the genre known as the critical 'Volksstück'. Rainer Werner Fassbinder (1946-82), later to gain international renown as a film director, adapted *Pioniere* for stage and television, claimed that he would never have become a writer unless he had read Fleißer's play, and dedicated his film *Katzelmacher* to her in 1970. Martin Sperr (born 1944), author of the seminal *Jagdszenen aus Niederbayern* (1966), claimed her as a direct antecedent, and Franz Xaver Kroetz (born 1946), the most widely performed living German playwright of the 1970s, wrote in her praise, holding her up as a countermodel to Brecht and instigating the collected edition of her works, the *Gesammelte Werke* of 1972. These three were to become known as 'die Söhne Fleißers', and it was they who put her back on the literary map. The 1970s witnessed a succession of Fleißer revivals, first of *Pioniere* (11 productions and 201 performances between 1970 and 1979), then of *Fegefeuer in Ingolstadt*. At the same time the publishers Suhrkamp released a collection of her stories, paving the way for renewed attention to her prose works, especially by feminist critics. By the time of her death in 1974, Fleißer had been rescued from obscurity. Although she has never appealed to a mass readership, she is today regarded as one of the outstanding and most original writers of her generation.

Pioniere in Ingolstadt

The genesis and the three versions: 'fast schon ein Auftrag von Brecht'[16]

The idea for *Pioniere in Ingolstadt* came from Brecht. During a visit to Augsburg in 1926, Fleißer told him of the recent arrival in Ingolstadt of a regiment of soldiers who were to construct a bridge across the 'Künette-graben'. She described it as an 'Invasion' (I, 442). Brecht, who was at the time himself working on a play on the theme of militarism and its effect on the individual, *Mann ist Mann*, was fascinated by the dramatic potential of the topic. He encouraged Fleißer to write a play about 'eine solche militärische Invasion in einer kleinen Stadt mit ihren Auswirkungen auf die Bevölkerung' (I, 442) and made precise suggestions about the form and content of the work as he conceived it. Despite great difficulties with the material, due largely to her ignorance about military matters, Fleißer persevered, producing a text with which she was not entirely happy. In this original form the play was produced in Dresden on 25 March 1928, proving a modest success. In the following year Brecht arranged its performance at the 'Theater am Schiffbauerdamm' in Berlin, where his own *Dreigroschenoper* had just achieved sensational success. In his desire to outrage the bourgeois theatregoing public and revolutionise the stage, Brecht demanded significant changes to the play as it had been performed in Dresden. He wished to sharpen its political (anti-military) message, but, above all, to radicalise its sexual theme by adding 'Pfeffer' (I, 445). On his instructions, Fleißer added a passage in which three schoolboys discuss the female anatomy, and he himself made directorial interventions to heighten the shock-effect of other sexual scenes. During rehearsals it became clear to Fleißer that Brecht had little respect for her text, and was concerned only to provoke a scandal: 'ein vorbereiteter Skandal mit politischem Hintergrund' (IV, 530). She withdrew her co-operation at the last moment, leaving Brecht a free hand. The first night, in the presence of the deputy police president, who himself had once been a 'Pionier' in Ingolstadt, lived up to Brecht's expectations. The play was immediately banned, pending alterations to the offensive scenes (which were rapidly made), and news of the sensation soon spread from Berlin to Ingolstadt, where Fleißer was denounced as a 'Nestbeschmutzerin'. The *Pioniere* scandal haunted her for years to come: it made an early return to her home-town difficult, and involved her in protracted legal action against the town's mayor, who had published a defamatory press statement about the affair. It strained her relations with her father and led, in the same year, to her rift with Brecht. Above all, it left her feeling exploited and humiliated.

Fleißer left the play untouched for many years, apparently regarding it as an unfortunate episode of her youth.[17] It was Brecht's widow, Helene Weigel, who prompted her in 1967 to rework the text with a view to a possible performance by the 'Berliner Ensemble' in East Berlin, a performance which never actually materialised. Thus arose the definitive 1968 version of the play, premièred at the 'Residenz-Theater', Munich, on 1 March 1970, first published in book form in the *Gesammelte Werke* in 1972, and included in the Suhrkamp paperback *Ingolstädter Stücke* (1977). Quite independently of Weigel's interest in the play, attention had been drawn to it by Rainer Werner Fassbinder. He produced (originally against Fleißer's wishes), an adaptation of the 1929 version under the title *Zum Beispiel Ingolstadt* at the Büchner-Theater, Munich, in 1968, and made a television version in 1971. The Fleißer revival of the 1970s can actually be traced back to the rediscovery of *Pioniere* by Fassbinder. In his 1968 adaptation, Franz Xaver Kroetz played the role of the sergeant. Thus it was that Fleißer's influence spread.

Fleißer's 1968 version, reproduced in this volume, constitutes a considerable improvement on the earlier versions in terms of dramatic tautness, coherence, and clarity of theme. Paradoxically, only when finally freed from Brecht's influence (and interference) did Fleißer fully realise the Brechtian element of the play: 'Der gesellschaftskritische Einfluß Brechts auf mich kommt erst in meiner Bearbeitung von 1968 deutlich heraus' (I, 447). A distracting sub-plot involving the sale of a car was removed, and the episode concerning the theft of the wood (3. Bild, 5. Bild) was added in order to reinforce the tensions between the military and civilian population. The play is almost twice as long. The characterisation of Unertl, Fabian, Berta and Alma was sharpened in order to show more clearly the influence of their social situation, and the brutalisation of the soldiers by the military system was made more explicit. Although she left the historical setting of the play in the year 1926 unaltered, Fleißer sought to achieve universal relevance through her exposure of the enduring mechanisms of oppression evident in bourgeois society.

The characters: 'ein Stück über Soldaten und Mädchen' (I, 442)

Brecht's intention in suggesting a play dealing with the military had been primarily political. Despite the dissolution of the Imperial Army in 1919 as a result of the Treaty of Versailles, the military ethos, as an integral part of both Prussian and Bavarian culture, retained enormous prestige and respect. To attack the values behind such militarism, as several writers did at the time, was to strike at the very heart of the Establishment. Brecht's own play, *Mann ist*

Mann, was originally conceived as a critique of the dehumanising effects of military life on the individual, indeed, as a study of the complete malleability of the individual in the military collective. An early experiment in epic theatre, this play is a demonstration of an idea through a somewhat fantastic plot in which characters are reduced to mere types. The drama has little realism or human interest, and works like a comic parable. Fleißer, however, as outlined above, was more interested in tangible experience and psychology than in overtly political metaphor: 'ich habe nicht so politisch gedacht wie Brecht', she later pointed out (I, 442). The political critique in her works is never explicitly formulated, though it is undoubtedly implied.[18] In *Pioniere*, as elsewhere, Fleißer approaches her material through the everyday activities of ordinary people presented through a simple plot, placing special emphasis on relations between the sexes, once more on 'etwas zwischen Männern und Frauen': 'Das Stück spielte zwischen Soldaten, Dienstmädchen und einem Bürgersohn. Die Handlung war in bewußt alltäglichen Zügen zwischen Einmarsch und Ausmarsch von Pionieren in einer Kleinstadt verlegt' (IV, 476). The aspirations and frustrations of young people stand in the foreground. An analysis of these is able to reveal the broader social determinants, and thus the sociopolitical context, of what is ostensibly an unpolitical play. Through her exploration of the processes of socialisation and maturation experienced by her characters, Fleißer succeeds in imbuing the individual case with a more general validity.

The circumstances of Fleißer's characters in *Pioniere in Ingolstadt* are highly reminiscent of those in her stories, and in her earlier play *Fegefeuer in Ingolstadt*. The young women and men are engaged in a search for some form of personal fulfilment or release. At every turn, though, they are hampered by the distorted values instilled in them by their elders and peers, and are subject to the hostile and brutalising pressures exerted on them by society at large. The quest for fulfilment reveals itself, as ever, in the intensely private sphere of love: *Pioniere* might well be considered, on one level, as an exploration of love in a variety of its manifestations. Its romantic interest, if it might be called such, revolves around the shifting relations and interrelations between the four characters Berta, Alma, Korl and Fabian. Berta, pursued by Fabian, falls in love with the soldier Korl, who enjoys a casual encounter with Alma and has a number of other liaisons elsewhere. Alma, Berta's best friend, experiments with prostitution, but ultimately ends up with Fabian, who in turn renounces any hope of winning Berta. Berta fails to win Korl's lasting interest and remains alone. The conceptions of love illustrated by these characters range from the romantically idealistic to the unashamedly carnal. In their interaction an entire spectrum of attitudes and behavioural patterns emerges.

The relationship between Berta and Korl is, perhaps, the central thread running through a play which otherwise has little strict unity in the classical sense. Berta, the idealist, is driven by notions of self-realisation through self-abandonment to love. Like Alma a 'Dienstmädchen', she is hesitant about the apparently quite natural option of a relationship with her employer's son, Fabian, who is actively pursuing her. This relationship has the clear advantage of convenience and might possibly guarantee her security. Interestingly, Fleißer's own father married his housemaid in 1923, after his first wife, Marieluise's mother, had died in 1918. But Berta insists instead on her right as an autonomous individual to exercise freedom of choice ('aussuchen kann ich ihn mir schon', 38). She lacks the coquettish self-assurance of Alma and is clearly concerned to protect her purity and reputation. Her conception of love is a romantic one, which, as her occcasional use of sentimental clichés demonstrates, she has presumably gleaned from popular films and cheap magazines. It is her misfortune to fall for the soldier Korl, who is initially charming and chivalrous, but soon displays his complete indifference to any emotional attachment and openly declares his exclusive interest in sexual relations. 1. Bild/4 is a skilful dramatisation of the first rendezvous between two young people who approach their relationship with vastly different expectations. Berta, no doubt as a result of her upbringing, is coy, almost touching in her naive expressions of sincerity and cautious attempts at intimacy. She is sensitive, potentially jealous and possessive, and clearly in search of genuine contact. In the vain hope that she might be Korl's only love, she insists on her own individuality ('ich bin nicht wie die anderen', 43), and is proud of the fact that she has saved herself for a lasting relationship. For her the kind of intimacy which she shares with Korl can ultimately have only one natural outcome: marriage (75).

Korl, though, refuses to acknowledge, let alone negotiate on, the binding terms which Berta would like to set for their relationship. Highly promiscuous himself, he naturally assumes the same of her, indicating that infidelity, manipulation and emotional indifference are the conventions within which such encounters typically take place. And it is precisely as typical, as a normal, healthy male (in his terms), that Korl would like to be understood: 'ich bin wie alle anderen' (43). For him sexual contact is the natural, and mutually anticipated, outcome of any such meeting between man and woman ('Warum, meinst, bin ich mit dir gangen?', 44), and he soon seeks to assert his male domination through the threat of physical violence. In the face of Berta's continued resistance he adopts a different tactic, presenting her instead with an ultimatum ('Wenn ein Mädel nicht zieht, tu ich nicht lang um', 44), which is intended to force her acquiescence if she wishes to maintain their relationship. Berta is finally 'in die Enge getrieben' (44), and clearly loses this first encounter. Given Korl's stated antipathy to any form of romantic involvement, she is unable to apply any pressure in pursuit of her interests.

This scene, much admired for its linguistic compression,[19] sets the framework for the love theme of the play as a whole. In spite of Korl's consistent rejection of anything other than a casual sexual adventure, and despite evidence that he has liaisons elsewhere (10. Bild/2), Berta continues to pursue her dream of profound emotional commitment. Korl thwarts it at every stage, explicitly warning her that he is repelled by her romanticism and that her devotion will bring her only suffering. Suffering, though, is an integral part of woman's service to man as Berta understands it ('ich will leiden', 58). Under the pressure of her continued refusal to take his declarations of indifference seriously, and of her insistence that deep down he must love her too, Korl openly displays his aggression towards the opposite sex: 'Einen Fetzen muß man aus euch machen' (75). To the bitter end Berta retains her conviction that her sincerity will reap its own reward ('Mich wird einer noch brauchen, das weiß ich', 76) and that Korl will ultimately relent. He does not. She finally surrenders herself to him (14. Bild), and he quickly and unceremoniously deflowers her in the bushes, to the amusement of his colleagues. Disillusioned and resigned, Berta is left with the realisation that she has sacrificed her closely guarded virginity not to true and reciprocal love, but to an act of empty lust. Her romantic view of sexual union as the ultimate expression of mutual and binding love is destroyed.

The characteristics of the relationship between the sexes as illustrated here are aggression, conflict and open struggle. Berta's faith in her own value as an individual with something valuable to offer, indeed, her readiness to give herself even to the point of complete self-effacement, is crushed by the brutal cynicism of Korl. He tolerates no illusions and sees male–female relations as dictated by the forces of the market-place, with their laws of supply and demand: 'Heut muß ein Mädel sich was gefallen lassen, weil es zuwenig Mannsbilder gibt' (74). In this scenario the woman's role is passive, unquestioning acquiescence, while the man's is self-assertion. Korl, as a soldier whose way of life offers much opportunity for casual relationships and little scope for anything more lasting, has perfected his role as the unfeeling machine and successful seducer for whom emotions are utterly irrelevant:

BERTA War das alles?
KORL Warum? Was hat dir gefehlt?
BERTA Wir haben was ausgelassen, was wichtig ist. Die Liebe haben wir ausgelassen.
KORL Eine Liebe muß keine dabei sein. (90)

A whole series of scenes (1. Bild/3; 4. Bild; 10. Bild/2 and 3; 14. Bild) demonstrates that Korl is merely typical of the military mentality: unattached physical gratification offers a pleasant (and necessary) distraction from the pressures of regimental duties. Life off-duty is a search for sexual adventure, and there is intense competition for the favours of the local girls, involving a battle of wits and practised courtship rituals. 9. Bild shows that virtually all of the soldiers have found willing partners among the female Ingolstadt population.

But the notion of intersexual relations as a bitter conflict is by no means restricted to the military. The play traces not only the progress of Berta's experiences with Korl, but also, as a parallel development, the efforts of the similarly inexperienced Fabian to find a partner. The opening two scenes of the play juxtapose these two innocents seeking advice from their more experienced peers. While Berta in 1. Bild/1 hopes for a partnership which will allow her to retain her identity and dignity ('So wie du möcht ich nicht sein', she says to Alma, 38), the equally naive Fabian in 1. Bild/2 hopes for something much less distinctive: he simply wants to be like other men. He is, however, just as conscious as Berta of the dangers of entanglement with the opposite sex. His friend Zeck, ten years older and (so he would have us believe) an expert womaniser, warns Fabian in worldly-wise tones about the pitfalls of romantic encounters: 'Hauptsächlich sind es in der Liebe die Fallen, wo die bewußten Fußangeln lauern' (39). Confirming the experience of a number of characters in Fleißer's stories of the same period, Zeck identifies the ground-rules of male–female relations as 'die nackte Notwehr'(39). He echoes Korl's philosophy of the need to keep one's emotions well under control so as not to expose oneself to the dangers of involvement. Love is a battle for supremacy in which there can be only one victor: 'Du oder ich' (39). Like Berta, then (they are both seventeen), Fabian is to be initiated into the rituals of adult behaviour. Eventually, after resigning himself to the loss of Berta (10. Bild/6), in whom he shows more than a merely physical interest, Fabian does learn the rules. In the second 'Luitpoldpark' scene (13. Bild) he has drawn the conclusions from his fruitless quest for reciprocation. He will never again manoeuvre himself into a situation where he is the weaker: 'Wie du mir, so ich dir .... Einer will dem anderen Herr werden. Wer wird wen fressen?' (87). He goes into the bushes with Alma for his first sexual experience.

Having begun the play with the intention of winning the fundamentally sincere Berta, Fabian finally enters into a liaison with Alma. She, of course, stands from the outset of the action in stark contrast to Berta, and to an extent reminds us of Korl. Her independence is manifested in the fact that she no longer works as a 'Dienstmädchen' and is convinced that she will be able to earn a living from the soldiers (1. Bild/1). Her prostitution is a frank admission that she has divorced sexuality from any notion of love and regards it as a saleable commodity. Superficially she seems well suited to this role: she is self-assured,quick-witted and ruthless. In her vulgar exchanges with the soldiers she gives as good as she gets. To this extent she exhibits the emancipatory tendencies apparent to varying degrees in the heroines of *Die Ziege*, *Eine Zierde für den Verein*, and *Der starke Stamm*, and is ostracised by the local girls in 10. Bild/4 for stepping beyond the bounds of acceptability. But Alma is a less accomplished prostitute than she pretends, allowing herself to be outwitted by the 'Feldwebel' over an outstanding debt. More significantly, she is disillusioned by the whole love trade, in which men seek only to gratify their carnal desires. This should have been obvious to her from the beginning, but her experiences have brought home to her with great force the fact that she will not gratify her deeper aspirations in this way. Alma, for all her brazen coquettishness, has, like Berta, an ingrained desire for a different and better life. It is symbolised, typically for Fleißer, by her desire to taste city life (78). Her attempt at self-liberation through emotional and financial independence has failed ('Ich habe mich ins Freie gewagt, aber dort war es nicht frei', 77), but she continues to retain faith in her right to some form of fulfilment, for which (again like Berta) she uses the word 'heben' (78: cf. Berta on 75). Ultimately she arrives at more or less the same conclusion as Berta: men are unable to satisfy the deeper aspirations which women cherish. Her final, resigned liaison with Fabian is based on her desire to be his first experience, for once not to have to accept second-hand wares. It is a form of gratification as empty as the photograph which is all that Berta is able to retrieve from her dreams of a lasting relationship with Korl.

The basic truth that emerges from the various configurations of characters in the play, and towards which Berta and Fabian are educated, is that the game of male–female relations is truly a battle for survival in which one struggles to salvage some form of dignity. It is, indeed, 'die nackte Notwehr'. Each of the principal characters is brought to a stark realisation of the brutish facts which Korl and his military colleagues have long accepted. All of them pick out a personal path through the difficult and dangerous labyrinth of relations, and respond in their individual way to the experiences which engulf them.

Society and the individual: 'der Druck geht nach unten' (55)

For all the intimacy of their private experiences, the characters of *Pioniere in Ingolstadt* are not divorced from the wider reality of the world. On the contrary, Fleißer is concerned at every turn to suggest forces beyond their control which shape them, conditioning their attitudes and behaviour. Her characters are simultaneously individual and typical: individual in the sense that they are endowed with intense inner lives, typical in as far as their relations with each other follow certain predictable patterns within a rigid system of dependencies and pressures. Although the play attaches too much importance to individual psychology to function as a parable, its characters are selected with a view to their representative status. Thus, according to Fleißer, Berta and Alma embody two opposing conceptions of love: 'Mit Berta und Alma habe ich die für mich faßbare Spannweite von jungen Mädchen dargestellt' (I, 442). When Brecht suggested the idea of the play to Fleißer, he sketched out his ideas for characters purely in terms of their social and professional role: 'Es muß ein Vater und ein Sohn hinein, es muß ein Dienstmädchen hinein ... Die Soldaten müssen mit den Mädchen spazierengehn, ein Feldwebel muß sie schikanieren' (I, 442). Fleißer, with her customary interest in realistic observation, added much flesh to the bones of Brecht's conception, but the typicality of her characters was preserved. This much she did learn from Brecht, claiming years later: 'Ich schildere aber keine bestimmten Figuren – niemals – sondern immer nur Typen. Ich strebe das Typische an' (*Mat.*, 344). Indeed, on a deeper level the entire plot and setting of the play indicate a wider social–political context. As Fleißer frequently pointed out, the Ingolstadt of the title is intended not as the specific or exclusive location of the play, but as a paradigm of provincial life: 'Aber selbst wenn meine Stücke *Pioniere in Ingolstadt* oder *Fegerfeuer in Ingolstadt* heißen, so ist damit nur eine gesellschaftliche Lebensform gemeint. Ingolstadt steht für viele Städte'(*Mat.*, 352/53. See also *Mat.* 348 and I, 447).

The way in which the play illuminates individual attitudes and actions as a projection of general conditions is most evident in its military theme. Brecht conceived the play originally as a study of military behaviour in a provincial town, and advised Fleißer to mix with soldiers in order to gain an insight into their mentality and language. As Fleißer intended it, the play was not to be a critique of the military as such, but rather about the problems of military life: 'Das Stück behandelt die Probleme der kleinen Soldaten und will kein Stück gegen das Militär sein, sondern gegen Mißstände beim Militär' (I, 447). Thus a central concern is the demonstration of the repressive hierarchy of the military regime, which treats individuals with extreme brutality and leads them to treat others in much the same way. Seen in this light, Korl is simply a product of the system: 'Männer beim Barras sind immer Männer unter Zwang' (I, 447). Like his fellow soldiers, he is constantly exposed to the harassment and aggression of his immediate superior, the 'Feldwebel', who drives him to work ever harder (8. Bild), humiliates him in front of Berta in order to demonstrate his own power (4. Bild), and treats him and his colleagues throughout with dictatorial sadism. The attempt on the sergeant's life (6. Bild) and the refusal to rescue him from drowning (12. Bild) are expressions of the hopelessness – Fleißer uses the word 'Ausweglosigkeit'(IV, 51) – engendered in the soldiers by a rigidly structured and authoritarian hierarchy. In this system, individuals are reduced to functions, and are quite replaceable. Thus in 14. Bild the new sergeant is simply 'eine neue Auflage des ertrunkenen Feldwebels' (92). In the normal course of life off-duty, a 'Feldwebel' would not even the acknowledge the existence of a 'Pionier' (56): in this hierarchy, the ordinary soldier is the lowest point on the scale.

Military life is presented as the very epitome of unfreedom: 'Für uns gibts keine Freiheit', says Münsterer (79). Fleißer is clearly concerned to show, though, that exploitation and bullying are not inflicted only on the lower ranks. The 'Feldwebel', too, is under pressure, and responds with his sadistic treatment of the 'Pioniere' to the harassment imposed upon himself from above. He correctly assesses the mechanisms of oppression, which always hit the weakest hardest:

FELDWEBEL ...In solchen Fällen wird der General ein Stier, und der Major wird ein Stier, und der Hauptmann wird ein noch größerer Stier. Je mehr nach unten, desto reißender der Zorn, und desto mehr wirkt es sich aus. Der Druck geht nach unten.
FABIAN Was machst du damit?
FELDWEBEL Ich gebe ihn weiter, den Druck.
FABIAN Siehst du!
FELDWEBEL Und ich werde einen Schuldigen finden, und wenn er nicht schuldig ist, dann mache ich ihn schuldig. (55)

Just as the sergeant vents his frustration on the common soldiers, and compensates for his own weakness by cheating Alma (10. Bild/1), so, too, do the common soldiers seek an outlet for their frustration and degradation in their treatment of those weaker than themselves: the women. This is the mitigating explanation for Korl's aggressively non-committal attitude to Berta. At the end of a day of hard labour and systematic harassment, he needs to let off steam: 'Den ganzen Tag muß ich mich schikanieren lassen, bei den Weibern lasse ich mich aus' (75). What he does not need is the extension of pressure and obligation into his private life. For the soldiers, then, casual relations with girls represent freedom and distraction from the otherwise intolerable pressures of regimental duty. The local girls are freely interchangeable with others; they have no distinct identity or value from the point of view of the soldiers, who simply move on to the next town at the end of the play. This is the only freedom the soldiers know. When Berta asks Korl if there is anything she can do for him, he simply answers: 'Mich einen freien Mann sein lassen' (75: cf. 'Mich muß man laufen lassen', 58, echoed by Münsterer, 79). Hence his repeated insistence on his need to be allowed the scope to relax and unwind without ties. Hence, also, his need to assert his domination over the devoted Berta, his determination to dictate the terms of their relationship, and his celebration of his power over her (10. Bild/5). For Korl, dreams of personal fulfilment are a thing of the past (78).

For the 1968 revision of the play, significant alterations were made above all to the scenes depicting civilian life. Fleißer herself pointed out that she had been at pains in her reworking of the text to intensify its sociocritical impact: 'Ich habe alles vertiefen und anzureichern versucht mit Gesellschaftskritik' (I, 447; see also *Mat.*, 348). Among the most important modifications was the development of the scenes in the 'Haushalt Unertl' (2. Bild, 7. Bild), where an insight is given into lower middle-class ideology, especially with a view to love and marriage and the treatment of servants. It becomes clear here that the authoritarianism evident in the military finds an uncannily precise parallel in the patriarchal structures of society as a whole. The play thus became an exploration not just of the military itself, but of militarism as a social phenomenon (see *Arbeitsteil* 2). Unertl is the ruthlessly exploitative self-made man , 'der ewige Spießer' (*Mat.*, 348), who expects absolute subservience from those in his employment. In his bullying of Berta ('Bei mir daheim will ich mich pelzen', 45), which closely resembles the behaviour of the sergeant towards Korl, he confirms the fundamental truth of human relations: 'der Druck geht nach unten'. He is a parody of the domestic tyrant, and would be an almost comic character were it not for the sinister implications of his attitudes. Quite apart from the demands he makes as the employer of Berta, the sexual rights over her which he claims for his son, and the shocking insults he delivers in 7. Bild, his view of interpersonal relations stamps him as an utterly chauvinistic boor. His age and status dictate that he should remarry, yet he can conceive of wedlock only as a financial arrangement which has not only a credit side ('Ich heirate eine Frau, die mich ersetzt in meinem Geschäft, und sonst nicht', 46) but also a debit side ('Bin ich verheiratet, dann hat die Mark fünfzig Pfennig', 46). No wonder, then, that he has been rejected in his suit for the hand of his shop-assistant. Yet Unertl, too, no doubt, is as much a victim as the other aggressors in the play. He is presumably subject to the same market forces which prompted Bepp Haindl, in breach of an earlier agreement, to force his new wife Marieluise Fleißer to work in the family business following their marriage in 1935. Like so many of Fleißer's characters, the frustrated Unertl hopes for realms of experience which are closed to him: he, too, uses the word 'heben' (46).

The more closely one looks at the play, the clearer it becomes that none of the characters has much freedom to manoeuvre, nor any real chance of escape. Rather, they are hemmed in on all sides by authoritarian structures (military or domestic), or by expectations of conformism which manifest themselves particularly clearly in peer-group pressures. In the play as a whole the women are certainly less free than the men. Berta, for example, is the victim of aggression from both Korl and Unertl, who, in different ways, assert their male rights over her. Here Fleißer's abiding concern with forms of female oppression is evident: the play identifies the kind of pressures exerted by men on the woman's private sphere which are more explicitly analysed in the essay 'Jahrhundert – gedrittelt' of 1932/33. There are abundant signs of proto-feminism here. But such pressures do not emanate in *Pioniere* (or elsewhere in Fleißer) simply from the opposite sex; they are also evident within the sexes. Thus Fabian is presented as a member of a group of men who advise and cajole him with regard to his sexual initiation (the 'Männergespräch', 38 f.), even demanding visible tokens of his conquest (48). The pressure to succeed and conform is considerable. And Alma, though far more independent, is forced to defend her behaviour before two girls who accuse and abuse her (10. Bild/4), offering society's wider view of her prostitution. In the scenes with Berta, meanwhile, she assumes a similar role to that of Zeck *vis-à-vis* Fabian, offering advice and making it clear that one's normality depends on success with the opposite sex (37). The acquisition of a partner is regarded universally in this play as the prerequisite for social acceptability.

The final manifestation of tension and conflict evident in the social structure of the play is that between the military and the local population. This theme was intensified for the 1968 revision by the addition of the sub-plot involving the theft of the wood by the 'Schwimmverein' (3. Bild, 5. Bild, with repercussions in 4. Bild and 11. Bild). The hostility between the two groups revolves initially around their rivalry for the affections of the local girls (Fabian *v.* Korl). But there are more fundamental tensions than this, springing from the fact that the civilians resent the presence of the military (2. Bild, 3. Bild), while the soldiers vent their frustration with the restrictive military life in aggression towards the local men (11. Bild). The theft of the wood by the swimmers, in fact, serves to amplify the theme 'der Druck geht nach unten', exposing the mechanisms of descending harassment with great clarity: the buck passes from general to major, to captain, to sergeant and on to the pioneers. The latter, as outlined above, use the girls as an outlet, but also seek compensation in their bullying of Fabian in the so-called 'Tonnenszene' (11. Bild). Their pent-up frustration with the regime is evident throughout the play (in 4. Bild, 6. Bild, 8. Bild, 10. Bild/7, 12. Bild), but it finds its clearest expression in their treatment of Fabian, whom they constantly deride as 'Zivilist':

ROSSKOPF ... Du bist auch so einer, den es nichts angeht. Aber es geht dich was an. Zum Teufel! An der Eskalierwand geschunden, mit voller Ausrüstung und Gepäck durchs Gelände gehetzt.
MÜNSTERER – ganze Kilometer durch nasse Äcker und Straßen gekrochen.
ROSSKOPF – und das alles geht dich nichts an. Aber das wär grad für dich was, und das lasse ich jetzt an dir aus. (82)

The contempt which Fabian feels at the brutality and inhumanity of the soldiers, indicative of the wider resentments harboured by civilians against the military, is voiced in 12. Bild, where he witnesses the refusal of the 'Pioniere' to rescue the sergeant.

At all levels of the play, then, private actions are generalised as a reflex to broader social conditions. The characters, alienated from themselves and from each other, lack the critical distance which would enable them to identify the social norms which determine their lives.

*Historical and contemporary relevance: 'es gibt Stücke, die kann man immer spielen' (*Mat., *348)*

Pioniere in Ingolstadt is a play firmly rooted in its historical setting. Beneath the list of *dramatis personae* we read 'Die Komödie spielt 1926', i.e. the action is set at the time the play was written. Brecht, perhaps in order to heighten the text's provocative impact, moved it back some fifteen years into the glorious days of the 'Kaiserzeit' for his 1929 production. Fleißer regarded this idea as 'Unsinn' (I, 446). For her the play was instantly recognisable as a product of, and a reaction to, the conditions of the Weimar Republic, and despite her concern to intensify its modern relevance for her 1968 revision, she never had any doubt that its original setting should remain untouched. Indeed, she did not regard this setting, nor the fact that conditions had changed since, as an impediment to contemporary, and even lasting, validity:

> Die Zeit habe ich lassen müssen, das Stück spielt 1926. Es sind heute nicht mehr dieselben Pioniere, die Pioniere sind heut technisch Verbeamtete von einem hohen Ausbildungs- und Wissensgrad, sie haben einen viel höheren Sold. Und solche Dienstmädchen gibt's heute auch nicht mehr, damals gab es sie noch ... Also das Jahr 26 muß ich schon lassen, ich sehe auch nicht warum nicht. Es gibt Stücke, die kann man immer spielen, die sind trotzdem aktuell (*Mat.*, 348).

In the 1968 version, then, the historical, and the geographical, location remain highly specific. Quite apart from the title, specific place-names ('Altwasser', 38; 'Unterer Graben', 58; 'Goldknopfgasse', 61; 'Kaiserwall', 74; 'Luitpoldpark', 10. and 13. Bild) identify the town in question as Ingolstadt. The situation of the Bavarian provincial town between the wars, deprived of the large military presence which had guaranteed its relative economic prosperity since 1870, and which was therefore both unsettled and excited by the arrival of the Küstrin soldiers in 1926, is clearly reflected. Indeed, the rather claustrophobic, Catholic–conservative atmosphere of the Bavarian backwater is impressively evoked within these few pages, and the consistent use of regional language adds further to the specific character of the text. In this respect, *Pioniere* constitutes another chapter in Fleißer's persistent exploration of the attitudes of provincial people, based very much on her own experiences.

Quite apart from its specific local flavour, the play captures something of the social and political conditions of Germany in the 1920s. Following the capitulation of 1918 and the subsequent dissolution of the German Empire,the military, and militarism, remained a burning and emotive issue. The (Prussian) military ethos continued to be an important part of German culture and of the popular consciousness, despite the fact that the total strength of the 'Reichswehr' had been fixed at only 100,000 men by the Treaty of Versailles. Nationalist and conservative groups kept the glorious military ideal alive through their propagation of the notion that the Imperial army had remained undefeated in battle and had been 'stabbed in the back' by socialist and pacifist groups. During the troubled beginnings of the Weimar Republic, marked by instability and political extremism, the army was an important instrument of social order and a significant political factor, which shortly afterwards was to be exploited by the National Socialists in their unbridled anti-republicanism and xenophobia. The Nazis were able to direct the frustrations of the people with the unpopular Weimar parliamentary system into anti-democratic channels, glorifying the army, and even war: notions of discipline, order, hierarchy were characteristic of an entire ideology. Thus, to the politically progressive, the military represented a bastion of conservative values, and was synonymous with political reaction. Brecht, then, was right to see an assault on the authoritarian and brutalising structures of the army as a political provocation of the conservative Establishment. Fleißer always denied that her play was intended as a critique of the military *per se*, and was taken aback by the vehemence of the reactions it aroused. It does, however, place the army in an unflattering light, and provides, in the closing speech of the new 'Feldwebel', an ironic contrast between the army's view of itself as strictly disciplined 'Staatsbürger in Uniform' (93) and the reality of the soldiers' behaviour. Its cherishment of the halcyon days of the great Wilhelminian Empire (1871-1919) is indicated in 4. Bild, where the colours of the Empire ('Schwarz–Weiß–Rot', abolished in favour of 'Schwarz–Rot–Gold' in 1919) are celebrated in the musical accompaniment to the scene.

The play also suggests, in the Unertl and 'Schwimmverein' scenes, something of the anti-democratic mentality which allowed, and even encouraged, the growth of fascism at about this time, especially in Bavaria (in 1926 Hitler, released from prison, was developing his NSDAP from his Munich headquarters). The chauvinism of Unertl, the secrecy and dishonesty of the swimmers, with their pronounced group identity ('Vereinsmentalität') and belief that their petty criminal activity is some form of virility test, indicate underlying civilian attitudes which were later to be transformed into blatant elitism, nationalism and racism. Fleißer's novel, *Eine Zierde für den Verein*, sketches in more detail the provincial attitudes which proved such a fertile breeding ground for fascist ideas.[20] It closes with a scene which suggests the violence inherent in society at the time: the depiction of a bloody brawl between two groups of men (one group, incidentally, being a 'Schwimmverein'). Furthermore, *Pioniere* allows a glimpse of the economic situation prevalent at the time, so conducive to the spread of anti-democratic values. An atmosphere of provincial backwardness and insecurity hangs over the drama. Despite the fact that the action is set in the so-called period of stabilisation of the Weimar Republic (1924-29), economic pressures are apparent beneath the surface. Girls like Berta accept the degradation of life as 'Dienstmädchen' for lack of more satisfying employment; Alma, having lost her position, will have difficulty in finding a new job (Berta: 'Was machst dann? Jetzt stehst da', 37), and is driven to prostitution to earn a living (58; 71: 'Was essen müssen ist ja noch keine Schande'). She dreams of escape to Berlin (78). Unertl is subject to the intense pressures of competition (2. Bild), and the town of Ingolstadt has insufficient funds to have its bridge constructed privately (38) or to subsidise its swimming club (49). The economic difficulties apparent in Ingolstadt at the time – in her novel Fleißer speaks of 'der entmilitarisierten Stadt mit nur neunundzwanzigtausend Einwohnern und zehn Prozent Arbeitslosen' (II, 9) – are at least latent in the play.

The outrage which Brecht's 1929 production of *Pioniere* provoked is, perhaps, the surest sign of all that Fleißer's play was sufficiently sensitive politically to touch a raw nerve among her contemporaries. The reaction of the right-wing Bavarian press, and even more so of newspapers sympathetic to the NSDAP, was violent in the extreme, confirming the suggestion in the play of proto-fascism (See *Arbeitsteil 2*). That the military should be so heinously maligned was bad enough. That a woman should be the perpetrator of such defamation (and obscenities) was unspeakable. The bigotry and venom which was later to prove so typical of the Nazi 'gleichgeschaltete Presse' is clearly evident here. A campaign of hatred was launched against the entire *Pioniere* phenomenon: the author was a woman, the *de facto* director a Marxist, the theatre owner (Ernst Josef Aufricht) a Jew, and the production had taken place in the decadent metropolis, Berlin.[21]

If *Pioniere* is so obviously a product of its time, to what extent is its relevance restricted to its time? This question has been at the heart of the critical debate surrounding the work since the first production of the revised version in 1970. One leading critic, Joachim Kaiser, was scathing in his condemnation of the play as a relic from a forgotten age, while another, Botho Strauß, subsequently himself to become one of Germany's most distinguished playwrights, defended its validity as an exploration of persistent ideologies.[22] Fleißer's own view was close to that of Strauß. For her, though conditions had undeniably changed, the authoritarian structures of social power, exploitation and brutality evident in the 1920s were still apparent in modern bourgeois society. An important statement on the play written in 1972 (first published in its entirety in the fourth volume of the *Gesammelte Werke* in 1989) reads:

Hier, was ich mir über die *Pioniere* denke:
Pioniere in Ingolstadt ist ein Stück über die Ausweglosigkeit der kleinen Leute.
Ausweglos sind die Dienstmädchen, sie können sich nicht gegen die Ausbeutung wehren. Ihr Ausweg zur Menschwerdung hin wäre die Liebe, aber die fängt erst gar nicht richtig an. Sie geraten an die Soldaten, welche selber den ganzen Tag unter Druck stehn und den Druck loszuwerden versuchen, indem sie ihn weitergeben. Ausweglos sind nämlich auch die Soldaten.
Wenn aber Mädchen wie Alma der Ausbeutung zu entschlüpfen glauben, ist dies nur eine Täuschung. Ihr vermeintlicher Ausweg wird sie foppen, er führt sie nur in die

Grube hinein.
Zwar gibt es diese Dienstmädchen heute nicht mehr. Ausbeutung aber wird immer da sein, denn es werden immer Abhängigkeitsverhältnisse sein. (IV, 517)

Here the author defines the essence of her play as the hopelessness, the lack of freedom, apparent not just in the life of the oppressed soldiers, but in every sphere of the social and private existence of ordinary people. The insecurities and complexes, rivalries and aggressions of the characters can be traced back in every case to their place in the profoundly authoritarian social order, which deprives them of any scope for fulfilment and self-realisation. The play is to be read not as an analysis of a bygone age, but as a paradigm of all forms of dependency relations, wherever and whenever they may exist. One of the earliest commentators on the play, the theatre critic Alfred Kerr, abstracted this general message from the play as early as 1929, seeing it less as an anti-militaristic piece than as a comment on the 'Reste der übertünchten Raubtierschaft im hiesig-heutigen Mittelalter; der Zurückgebliebenheit' (*Mat.*, 70). Just as Ingolstadt is to be understood as everytown, so, too,does the social model constructed in this play have a validity beyond its immediate setting (see *Mat.*, 348). To deny this, as Botho Strauß pointed out, is to misunderstand the nature of literary relevance, which depends on the transferability of specific historical experience to a wider context.

If *Pioniere* is of relevance today, then, it is as a model rather than in its concrete detail. The same is true, of course, of all Fleißer's work, much of which was written in the 1920s but retains a great resonance in the 1990s. Society, in Fleißer's view, has remained fundamentally flawed. In 1972, speaking about *Fegefeuer in Ingolstadt*, she asked:

Ist es nicht auch jetzt so, daß wir eine unerlöste Gesellschaft haben? Ich sehe eine weitverbreitete Kontaktlosigkeit, Süchtigkeit, blinde Aggression, Unterdrückungslust, ein Imponiergehabe von Gruppen und immer wieder das Rudelverhalten gegen Außenseiter. Wo bleibt die Humanität? (*Mat.*, 364)

Like the remainder of her *oeuvre*, *Pioniere* is a critical work without an explicit moral. Conditions and injustices are demonstrated in the play, but there is no overt didacticism, no suggestion of positive change or of utopian alternatives. The characters do not comprehend, let alone analyse their situation. Yet conditions in the play cry out for a remedy: the moral is there, but it is implied rather than openly expressed:

Ich habe natürlich von Brecht gelernt, daß Theater nicht unreflektiertes Vergnügen sein kann, sondern daß aus einem Stück eine Lehre gezogen werden soll. Die Lehre strebe ich nicht gezielt an, sie entsteht unwillkürlich, sie muß da sein wie gewachsen. (*Mat.*, 344)[23]

Language and the 'Volksstück': 'Die Einfachheit ist die Spitze von einem Eisberg' (IV, 513)

As befits its milieu, *Pioniere in Ingolstadt* is written in the colloquial language of ordinary, uneducated people, in the case of the Ingolstadt population in a diction coloured by elements of provincial Bavarian dialect, while the soldiers are from Küstrin in Brandenburg. The characteristics of this diction include the abbreviation of past participles (e. g. 'gangen' for 'gegangen', 37; 44; 'worden' for 'geworden', 37; 43); the omission of personal pronouns (e. g. 'wie alt bist jetzt?', 37; 'jetzt stehst da', 37); the use of 'tun' as an auxiliary verb ('baun tun sie die Pioniere', 38); double negatives ('von keinem Herrn rede ich nicht', 42); irregular word order ('Das darf mir nicht einreißen, das sich die ihren Feierabend macht und mich laßt sie hängen', 45); the use of regional words ('sich pelzen', 45; 'spannen', 72; 'mer', 40; 'gell', 42 passim); and a host of other isolated deviations from standard norms ('das gwöhnst', 43; 'am Hut stecken', 49). Throughout her life, Fleißer remained attracted by the expressive possibilities of regional speech, which she saw as 'in sich schöpferischer als Schriftdeutsch' (IV, 537), as less conscious, more spontaneous and more differentiated than the standard language (*Mat.*, 345, 352). She was, however, equally aware of its limitations in terms of universal accessibility (*Mat.*, 352).[24]

The regionally flavoured colloquialism of *Pioniere* is, of course, an important aspect of realism in the play, intensifying the local colour. But it is more than this. At the time of writing the play, Fleißer was influenced by Brecht's ideas on 'naive' or 'gestic' language as an alternative to a purely naturalistic transcription of the phonetic qualities of everyday speech: '"Sie müssen kindlich schreiben", sagte er zu mir, "auf die Naivität kommt es an, schreiben Sie ganz naiv"' (II, 299). Brecht encouraged what he saw as an important talent in Fleißer: to find a form of expression in which naturalness (not naturalism) and stylisation are held in balance, a form which appears artless and spontaneous, yet carries an almost poetic intensity and a (social) symbolic power not easily achieved in the uniform idiom of the standard language.[25] Brecht was impressed by the graphic and intuitive quality of Fleißer's language, whose simplicity, he thought, showed an inherently dramatic concreteness and at the same time indicated typical characteristics of the provincial mentality. Fleißer herself defined this technique as 'im Gegensatz zum natürlichen das naive Sehen' (*Mat.*, 170), resulting in a language 'in der das Wesentliche ausgedrückt ist, so daß sich nichts davon wegnehmen und nichts hinzufügen läßt' (II, 300).[26]

A prime characteristic of the dialogue in *Pioniere* is its terseness. The people in the play express themselves in short, unpolished utterances which lack all sophistication. This is the language of the inarticulate, a language often referred to by Basil Bernstein's (now somewhat controversial) term 'restricted code', in contrast to the 'elaborate code' of the educated middle class.[27] Its grammar is simple and, in terms of standard norms, often faulty (e. g. Alma's 'Wärst gangen damit, dann tätest es wissen', 37); its syntax is compound rather than complex, relying predominantly on short main clauses; its lexical range is limited. It is the speech of uneducated provincial people in their everyday dealings, shorn of variety and subtlety, direct and naive, at times simply clumsy. Nowhere in the abrupt exchanges between characters is there any really sustained statement or response. Instead, communication proceeds via a series of brief utterances. Characters seem almost trapped by their inability to reach each other through words. On the whole, the language of the play shows a complete lack of self-analysis or reflection. It is unconscious in a way in which the language of Brecht's plays is not, and betrays a fundamental inability to grasp the complex reality of the world outside.[28]

The characters' inability to articulate their situation, and, above all, their feelings, is evident throughout the play. Berta, in her dealings with Korl, simply lacks the words to express her thoughts and emotions ('ich muß denken, wie ich es sage', 42). Alma searches for an adequate means to voice her deeper aspirations (78). Others, for example Unertl, the 'Feldwebel', Fabian, Korl, frequently use language to give others, and themselves, the feeling that they are in control, whereas in fact they clearly are not. They employ speech which is not their own, but 'borrowed', resorting to stereotyped formulations and clichés as a substitute for a truly individual idiom. Berta, in her desire for intimacy with Korl, can only couch her attraction in words borrowed from cheap literature: 'Die Sterne scheinen darauf. In deine Augen scheinen sie auch' (59); Korl retaliates in impersonal, set phrases: 'Die Frau wird von mir am Boden zerstört, verstehst. Da kenn ich keinen Bahnhof' (58). Unertl hides his despotism behind inflated references to external authority:' Der Haushalt ist die gesündeste Abwechslung. Das ist statistisch erwiesen' (45). Fabian adopts a pose of self-assurance and virility: 'Sie, ich täte nicht auslassen bei einem Mann, wie ich bin' (52). Alma uses pretentious words she presumably does not fully understand: 'ich bin überhaupt eine mondäne Frau' (58), sometimes wrongly ('eventunell', 38). Everywhere language functions as a mechanism which obscures true thoughts and identity and renders real human contact impossible.

Close examination of the dialogue of the play reveals that the entire process of communication between individuals is flawed. There is no true verbal interaction between characters, and Fleißer skilfully constructs speech events in which exchanges of ideas and feelings invariably break down, or at least do not take place on shared premises. Individuals frequently 'talk past each other', exposing an unbridgeable gulf between their respective perceptions and intentions. The result is an exchange, but never an interchange, of words in fragmented and cramped utterances whose subtext often reveals more than their surface meaning. Omissions, silences, clichés, sudden changes of topic indicate the deformed processes of communication in which characters cannot, or will not, co-operate to facilitate meaningful dialogue. The 'negotiations' between Berta and Korl, as Donna L. Hoff-meister calls them in an excellent analysis of verbal interaction in this play, are the clearest example of abortive communication.[29] In 1. Bild/4, 4. Bild, and 10. Bild/5 Berta makes repeated, if cautious, attempts to engage Korl in a discussion of emotional involvement, while he parries with a whole series of strategies designed to frustrate her intention: refusal to accept her propositions or to co-operate on her terms, contradictions, changes of subject, even threats. He degrades her intimate conversation to a ritualised exchange of evasive statements and generalisations:

BERTA Dann kann ich dir gar nicht helfen?
KORL Nicht, wenn du dich anhängst.
BERTA Ich will dir nicht aufsitzen.
KORL Heut muß ein Mädel sich was gefallen lassen, weil es zuwenig Mannsbilder gibt.
KORL Es gibt mich und gibt dich.
KORL Dir fehlt bloß die übersicht. Nämlich der Frauenüberschuß, der ist zu groß.
BERTA Es gibt mich und gibt dich, und andere brauchen wir nicht.
KORL Du sollst mich nicht immer so lieben, das macht mich noch rasend.
BERTA Wenn ich nicht anders kann.
KORL Weil du nicht schlau bist.
BERTA Wie wird man schlau? (74)

Similar tensions arising from dialogue in which the participants do not respond fully to their interlocutors, but resort to ritual exchanges, are evident in 1. Bild/3, and 4. Bild. In the play as a whole characters are confined by their inability to etablish contact with each other beyond the level of everyday banalities. Only where it acts as a bond to underpin group identity (the swimmers in 3. Bild, the military in 11. Bild and 12. Bild) does language promote community and solidarity. Here, though, it is degraded to the recitation of practised formulas (48; 50; 86) which have little to do with genuine communication.

Linguistic restriction in the play functions as an index, perhaps as the ultimate manifestation, of defective social relations. As a product of milieu, it illuminates the social situation of the individuals in the play in their isolation and limitation, and thus provides a crucial element in the social–critical constitution of Fleißer's text. This is the language of confinement and non-awareness rather than of clarification. It is informed by questionable norms and values, betraying profound insecurity and alienation, ignorance and aggression. Alongside its acclaimed 'realism', it is also an idiom which demonstrates its own deformation (Kroetz's 'Ausstellungscharakter'), drawing attention to the mentality of which it is the verbal expression. It is not the individuals who use such language who are criticised in this play, but the social environment which conditions both the characters and their language.[30]

The theme of limited consciousness as evinced through restricted, and restricting, language places *Pioniere in Ingolstadt* firmly in the tradition of the critical 'Volksstück' initiated by Horváth in the 1930s and more recently revived by Sperr, Fassbinder, Kroetz and others. Fleißer herself labels the play a 'Komödie', reserving the epithet 'Volksstück' for her later play *Der starke Stamm*. But the play is comic only in a particular sense (the tragicomedy of human relations), and has much in common with the neo-realistic form of popular drama which focuses on the problems of uneducated *petit-bourgeois* people in a provincial domestic milieu, placing special emphasis on the dynamics of verbal interaction as an indicator of distorted values. It was precisely this quality which attracted the attention of 'die Söhne Fleißers' in the 1970s and ushered in the Fleißer renaissance which has been such an important chapter in the recent history of German theatre. Here was a model of social drama which dispensed with an explicitly critical perspective and analytical sophistication, apparently merely observing the mechanics of a backward and authoritarian social structure. Yet the simplicity of the play is deceptive: 'Die Einfachheit ist die Spitze von einem Eisberg' (IV, 513). Behind what seems to be a simple and detached description of the surface of provincial life there is hidden, precisely in the faulty linguistic processes, a powerful critique of a system which holds individuals in a state of such unfulfilment.

Dramatic form: 'es muß zusammengebastelt sein' (I, 442)

In terms of its structure, *Pioniere in Ingolstadt* is clearly not a play conceived along classical (Aristotelian) lines. The conventional unities of time, action and place, guarantors of compactness and cohesion in traditional dramatic models, are cast aside. Fleißer dispenses with traditional act and scene divisions and organises her action into a succession of fourteen 'Bilder', some of which are subdivided to produce a total of twenty-three episodes. The period of time spanned by the play is unspecified: in fact, there are no co-ordinates to indicate time relationships between scenes. The action of the play is fragmented into a number of plot strands (e. g. the Berta–Korl–Fabian plot; the wood theft; the Unertl scenes; the attempt on the sergeant's life) which run parallel to, and interrelate with, each other, but lack the compelling continuity and integration found in strictly controlled classical drama. Nine different locations are used. The play thus has an 'open' structure similar to that favoured by the writers of the 'Sturm und Drang', especially J. M. R. Lenz, and by those important precursors of modern dramatic techniques, Georg Büchner (1813-37) and Frank Wedekind (1864-1919), rather than a clearly organised 'closed' form.[31] It is hardly surprising that it was seen as diffuse by many contemporary commentators.[32]

The structural principle which underlies the play is, of course, a quite deliberate antithesis of classical models. While the latter are characterised by structural symmetry, economy and absolute clarity of focus on a limited number of characters and events, Fleißer's play is, in effect, a montage of loosely related episodes with little causal connection. Conventional drama depends on a linear progression from exposition, through developing action to catastrophe and denouement, a process accentuated by the careful arrangement of *interdependent* scenes. Fleißer, on the other hand, simply strings together a succession of apparently *independent* episodes set somewhere between the arrival of the soldiers at the beginning and their departure at the end. Her logic of construction is quite different, dispensing with carefully engineered cohesion. The scenes of the play could easily be rearranged, as, in fact, they were during revisions of the text, or, indeed, in some cases even removed (as was done on the instructions of the police in 1929), without significant detriment to the substance of the whole.

In all of this, the influence of Brecht can be felt. It was he who suggested the idea of the play to Fleißer at precisely the time he was experimenting with forms of 'anti-Aristotelian' drama which eventually crystallised into his famous theory and practice of 'epic theatre'. In his search for forms of dramatic expression appropriate to the conditions of the modern age, Brecht returned to and refined the techniques of Lenz, Büchner and Wedekind, promoting a mode of theatre no longer centred neatly on the fate of outstanding individuals, but able, precisely by virtue of its apparent formlessness, to capture the complex collective forces of the twentieth century. The montage principle was ideally suited to this aim, and Fleißer reports that Brecht originally conceived *Pioniere* in these very terms:

> Anregung von Brecht: das Stück muß keine richtige Handlung haben, es muß zusammengebastelt sein, wie gewisse Autos, die man in Paris herumfahren sieht, Autos im Eigenbau aus Teilen, die sich der Bastler zufällig zusammenholen konnte, aber es fahrt halt, es fahrt! (I,442)

It is evident from the example of Fleißer's previous play, *Fegefeuer in Ingolstadt* (originally written in 1924, revised and first performed in 1926, revised again 1970/71), that she was attracted to an open dramatic form from the very beginning of her career. This play was conceived quite independently of Brecht, although its form bears witness to the profound effect his early works had on her (she mentions *Im Dickicht der Städte, Baal* and *Leben Eduards des Zweiten von England* in her biographical notes). So similar was its form that it was suspected by at least one critic of actually being the work of Brecht under a pseudonym.[33] It has much in common with *Pioniere*. But it is *Pioniere* which is the most Brechtian of Fleißer's works, and Brecht repeatedly referred to it, even years later, as a model example of early epic theatre.[34] Brecht's theories of epic theatre at the time (1928/29) were still at a somewhat rudimentary stage. This was the time of his study of, and subsequent conversion to, Marxism, which was to give his conception of political theatre a specific direction and a clearly defined aim: the transformation of society according to socialist principles. The core of Brecht's thinking at this early stage, before the formulation of his ideas on alienation, was the notion that drama could be assembled ('montiert') from a series of basic components, producing a chronicle effect (narrative or 'epic' rather than dramatic) which illuminates broad social forces in a pseudo-scientific way and does not dwell on the fortunes of individual characters. It is in this sense that Fleißer's play is 'epic': its focus is on milieu as much as on character; its structure is simple ('naive') rather than sophisticated; it emphasises the typicality of individuals' behaviour, exploring the power of collective forces; and it observes and demonstrates patterns of human behaviour with obvious detachment. However, as outlined above, Fleißer, as a fundamentally unpolitical writer, does not infuse her play with the critical and reformatory impact which Brecht ultimately sought via epic theatre. Her play is expository rather than didactic.

From earlier comments on the themes and implications of *Pioniere in Ingolstadt* it will be clear just how appropriate the 'epic' structure of the text is to Fleißer's concerns. The open form allows the author to illuminate developing relations between individuals, the processes which characters undergo, and the forces to which they are exposed in a way which would be impossible within the self-enclosed structures of more traditional drama. The latter presupposes a logic and linearity of progression which are incompatible with Fleißer's panoramic grasp of social interaction within a broadly defined milieu. Her drama has no traditional conflict between protagonist and antagonist, manifest in action and counteraction which move inexorably towards a definite conclusion. Instead it presents a social tableau which demonstrates conditions and situations, and lacks dramatic suspense. It is open-ended, dispensing with finality. Her technique is not integrative, but cumulative. Thus, the first four 'Bilder' of the play initiate four separate lines of action without seeking to establish a compelling link between them: the 'love theme' (from four different angles) in 1. Bild; the Unertl plot in 2. Bild; the wood theft in 3. Bild; and the military theme in 4. Bild. Since individual scenes do not depend logically on one another, this episodic structure is quite possible. 1. Bild and 10. Bild, both subdivided into constituent parts, illustrate this principle, as do other scenes which are not explicitly subdivided. In 4. Bild, for example, the focus shifts frequently and abruptly to different groupings of characters on the stage. Furthermore, in Fleißer's dramatic scheme, different strands of the plot can be allowed to run alongside one another, emerging and converging at various virtually interchangeable points.

Thus the Berta–Korl plot is initiated in 1. Bild/4, picked up again in 4. Bild, and suspended until 10. Bild. Meanwhile, Berta is seen in relation to Alma (1. Bild/1 and 3), Unertl (2. Bild, 7. Bild) and Fabian (4. Bild), while Korl is involved in the attempt on the sergeant's life (6. Bild, 8. Bild), presented as part of the military unit (4. Bild, 6. Bild, 8. Bild, 12. Bild), and seen briefly in relation to Alma (10. Bild/7). There is no rigidly defined order in this assemblage of constituent parts, and similar networks of relations and dependencies are constructed for all characters in the play. Fleißer's structural logic, then, dispenses with absolute integration and relies for its (rather loose) unity on a collage of situations linked by a common concern: the 'punctual' rather than systematic illumination of characters' behaviour as a response to social conditions. The organisation of the play is geared not towards a final resolution of a specific conflict, to which all parts are subordinated, but rather towards the gradual unfolding of 'eine .... militärische Invasion in einer kleinen Stadt mit ihren Auswirkungen auf die Bevölkerung' (I, 442).

Notes to the Introduction

1 IV, 305. Quotations from Fleißer's works are drawn from the collected edition, *Gesammelte Werke*, ed. Günther Rühle, of which three volumes were published in 1972 (Frankfurt). A fourth volume, 'Aus dem Nachlaß', edited by Günther Rühle and Eva Pfister, was added in 1989. Subsequent references to this edition are given in the main text, citing volume and page number. References to the play *Pioniere in Ingolstadt* cite the page number of the present edition. References to the invaluable collection of documents *Materialien zum Leben und Schreiben der Marieluise Fleißer* (ed. G. Rühle, Frankfurt, 1973) are also given in the main text, as *Mat.* followed by page number.

2 Apart from the two plays *Fegefeuer in Ingolstadt* (original title: *Die Fußwaschung*) and *Pioniere in Ingolstadt*, now known as the 'Ingolstädter Stücke', Fleißer's first published collection of stories, *Ein Pfund Orangen*, bears the subtitle 'und neun andere Geschichten der Marieluise Fleißer aus Ingolstadt'. Her play *Der Tiefseefisch* (original version 1929) was to be given the title *Ehe in Ingolstadt* for the revised edition on which she was working at the time of her death. In 1930 she published a story entitled *Gassenbesen in Ingolstadt* in the *Berliner Tagesblatt*.

3 The preoccupation with the effects of a Catholic education is evident above all in the play *Fegefeuer in Ingolstadt* (first version 1924), in the stories *Moritat vom Institutsfräulein* (1926) and *Der Venusberg* (1966), and in the novel *Eine Zierde für den Verein* (first published under the title *Mehlreisende Frieda Geier* in 1931).

4 Fleißer's difficult, indeed impossible, relationship with Haindl forms the basis of her novel and of a whole series of stories and radio plays, and even of a draft film-script (first published in the fourth volume of *Gesammelte Werke* in 1989).

5 'Ich bin in die Grube gefallen und wurde in der Grube verbraucht', she later wrote of this period in her life (I, 453).

6 The original title of the story *Avantgarde* (1963), the fictional but authentic treatment of her relationship with Brecht, was *Das Trauma*.

7 Although Fleißer's work is highly autobiographical, the experiences on which her

texts are based do not find their way into her fiction completely untransformed:' Das kann sich sehr verwandeln, bis es zu einer Geschichte wird' (*Mat.*, 343). See also II, 307/08, where she says of *Avantgarde*: 'Manches ist darin anders und aus dem Ganzen der Lebenserfahrung strömt es einem hinein, das ist ein unwillkürlicher Prozeß.'

8 'Wenn ich einen Menschen lieb habe, muß ich ihm jenes vornehme Vorrecht einräumen, mich von ihm verwunden zu lassen und die Härte des Augenblicks durch meine freudige Hingabe in etwas verwandeln, was meine Seele liebenswerter, der Liebe würdiger macht' (IV, 53).

9 Donna L. Hoffmeister, 'Growing up female in the Weimar Republic: Women in seven stories by Marieluise Fleißer', *German Quarterly*, 56 (1983), 396-407, p. 406.

10 Günther Rühle, 'Leben und Schreiben der Marieluise Fleißer aus Ingolstadt', I, 7-60, p. 49.

11 Fleißer explores the problem of 'die persönliche Würde der schaffenden Frau' in an essay of 1933, 'Jahrhundert – gedrittelt', IV, 427-9. She sees women caught in a 'Zwickmühle', caused by their desire to combine their increasing independence with their duties as wife and mother. While women were gradually coming to terms with their new role, she believed, men were unable to readjust, continuing to assert outmoded patriarchal claims.

12 Letter to Lion Feuchtwanger, 1954 (*Mat.*, 215). Her reversion to Catholicism, particularly strong in the latter years of her life, is confirmed by her sister Ella (Tax, 111).

13 'Ich schreibe Leben – aus Betroffenheit heraus' (*Süddeutsche Zeitung*, 9 January 1974). This insistence on the author's involvement with his/her characters is an important theme in Fleißer's essay of 1927, 'Der Heinrich Kleist der Novellen' (rediscovered in 1976 and now published in IV, 403-7), one of her few sustained aesthetic pronouncements.

14 The balance of involvement and detachment in Fleißer's stories is explored by Donna L. Hoffmeister: see note 9.

15 Letter to K.-P. Wieland, 14 December 1973. Despite her preoccupation with the problems of women, for example, Fleißer does not adopt an explicitly feminist position, as she pointed out in an interview in 1973 with Günther Rühle ('Jene zwanziger Jahre'): 'Das ist unbewußt, ich meine, daß ich mir nun ausgesprochen blaustrümpfige Ziele gemacht hätte, davon kann gar keine Rede sein' (quoted by Tax, 49). The 'unconsciousness' of Fleißer's social analysis is a distinctive feature of her writing.

16 Günther Rühle, *Mat.*, 358.

17 Letter to Fritz Voitel, 26 December 1962.

18 'Sicher lassen sich meine Stücke gesellschaftlich interpretieren. Aber mich würde es langweilen, wenn ich politisch schreiben würde' (*Süddeutsche Zeitung*, 9 January 1974).

19 See, for example, Franz Xaver Kroetz's essay of 1971, 'Liegt die Dummheit auf der Hand?' (*Mat.*, 379-86) and Donna L. Hoffmeister's study *The Theater of Confinement – Language and Survival in the Milieu Plays of Marieluise Fleißer and Franz Xaver Kroetz* (Columbia, South Carolina, 1983).

20 Fleißer wrote later: 'Meine Gestalten aus dem Kleinbürgertum sind Vorläufer ... Sie veranschaulichen durch ihr Eingesperrtsein in Zwänge den geschichtlichen Prozeß, der im Faschismus mündet' (letter to Rainer Roth, 12 December 1972).

21 'Das Asphaltparkett von Berlin wiehert gegen die Provinz' (*Mat.*, 84). By the 1920s

Berlin, with over four million inhabitants, was the world's third largest city. As a cultural centre it rivalled Paris and London, becoming synonymous with the excitement and energy of the 'Golden Twenties'. Its attraction for Fleißer was obvious. For right-wing thinkers, however, it was decadent, communist, morally and politically 'progressive', in contrast to the decent traditionalism of the provinces. The demonisation of urban life was an important aspect of Nazi ideology.

22 Joachim Kaiser, *Mat.*, 239: Botho Strauß, 'Bürgerdämmerung auf der Bühne', *Theater heute*, 1970, number 4, 18 f. See *Arbeitsteil 2*.

23 In her essay 'Sportgeist und Zeitkunst' (written 1927) Fleißer emphasises the need for literature to go beyond mere naturalistic description and to provide at least the seed for change: 'Es genügt nicht mehr Zustände und ihre vermeintliche erschütternde Sinnlosigkeit im Weltganzen zu zeigen. Es gilt den Samen des Willens zu säen, der ein energisches, sich selber vortreibendes Geschlecht erweckt und sich daraus ein Gewissen macht' (II, 319/20).

24 Of all Fleißer's works, only *Der starke Stamm* was originally written and performed in true dialect, and this, too, was later recast into a diction more closely approximating to High German for performances outside Bavaria.

25 Brecht later pursued these ideas further, when formulating his demands for a new form of 'Volksstück': 'Tatsächlich kann ein Bedürfnis nach naivem, aber nicht primitivem, poetischem, aber nicht romantischem, wirklichkeitsnahem, aber nicht tagespolitischem Theater angenommen werden. Wie könnte so ein neues Volksstück aussehen? ... Von der größten Bedeutung ist es, einen Stil der Darstellung zu finden, der zugleich artistisch und natürlich ist' ('Anmerkungen zum Volksstück', 1940, *Gesammelte Werke*, Frankfurt, 1967, volume 17, 1163).

26 Several leading critics praised the effectiveness of the diction of the production of 1929. Walter Benjamin, for example, noted: 'Die Worte der Fleißer tragen erstaunlich viel. Sie haben das Gestische in der Sprache des Volkes, schöpferische Gewalt, die sich zu gleichen Teilen aus einem entschiedenen Ausdruckswillen und aus Verfehlen und Ausgleiten zusammensetzt ...' (*Schriften*, Frankfurt, 1972, volume 4, p. 1028). Herbert Ihering wrote: 'Gänge, Bewegungen, Pausen. Die Sätze kamen nicht im sprachlichen, sondern im mimischen Zusammenhang ...' (*Mat.*, 75). See also the comments of Kurth Pinthus, quoted in *Arbeitsteil 2*.

27 Basil Bernstein, *Class, Codes and Control* (London, 1971). Bernstein's thesis has been criticised on the grounds that it regards 'working-class speech' as cognitively and therefore qualitatively deficient, and overlooks its distinct expressive power.

28 Franz Xaver Kroetz writes: 'Weil Brechts Figuren so sprachgewandt sind, ist in seinen Stücken der Weg zur positiven Utopie, zur Revolution gangbar .... Es ist die Ehrlichkeit der Fleißer, die ihre Figuren sprach- und perspektivelos bleiben läßt' (*Mat.*, 382/83). See *Arbeitsteil 2*.

29 See note 19.

30 Hoffmeister (see note 19) writes: 'The realism of ... Fleißer compels the audience to equate the strict confinement of the dramatic frame with the sociological confinement of the characters' lives' (p. 151). Typically, Fleißer seems to have achieved this effect instinctively, rather than as a result of a conscious effort: 'Es war mir nicht intellektuell bewußt, daß ich schon allein durch die Sprache das System kritisierte, das diese Sprache nötig hat. Der Endeffekt ist aber gegeben' (letter to Rainer Roth, 12 December 1972).

31 The distinction between two basic forms of drama, the 'open' and the 'closed', is explored by Volker Klotz in his highly influential study *Geschlossene und offene*

Form im Drama (Munich, 1960). The distinction is rather rough and ready, but it is useful as a general characterisation of fundamental dramatic structures.

32 See the reviews of the Dresden and Berlin productions in *Materialien zum Leben und Schreiben der Marieluise Fleißer*, 51-89.

33 Alfred Kerr, *Mat.*, 37.

34 Bertolt Brecht, *Gesammelte Werke*, Frankfurt, 1967, volume 15, 150 and 348.

Pioniere in Ingolstadt
Komödie in vierzehn Bildern
(Fassung 1968)

PERSONEN: Alma · Berta · Fabian, *siebzehnjährig* · Zeck, *siebenundzwanzigjährig* · Bibrich, *junger Schreiner* · Unertl, *Vater von Fabian, Geschäftsmann* · Korl Lettner, Münsterer, Roßkopf, Jäger, Bunny, *Pioniere* · Feldwebel · Der nächste Feldwebel ·'Fotograf · Pioniere · Mädchen · Bürger · Polizist.
Die Komödie spielt 1926.

1. Bild
Nähe Stadttor

1

Einmarsch der Pioniere. Musik. Berta und Alma. Bürger.

BERTA Warum singen die nachher nicht, oh du schöner Westerwald?[1]

ALMA Wärst gangen damit, dann tätest es wissen. Sagen laßt du dir nichts. Wie alt bist jetzt? Ein Jahr ist schon hin.
Pioniere.

BERTA Alma, ich will mich nicht streiten. Alma, mit dir kann man nicht gehn.

ALMA So? Meinst du?

BERTA Hat dich deine Frau auch weglassen?

ALMA Die kann mich nicht mehr weglassen. Ich bin ohne Stellung.

BERTA Was machst dann? Jetzt stehst da.

ALMA Da ist mir nicht angst. Der Pionier ist im Land.

Pioniere. Achtung, Augen links!

BERTA Ich möcht nicht wissen, Alma, was noch aus dir wird.

ALMA Du wirst es nicht ausbaden.

BERTA Gleich hängt man drin.

ALMA Von mir bist du noch nicht schlecht worden. Das kannst du nicht sagen.

BERTA Du bist die erste und laßt mich im Stich. Wenn die andern was über mich wissen.

ALMA Berta, nie! Weil ich das nie nicht täte.

BERTA Alma, das machen wir noch aus. Die Freundschaft muß bleiben.

ALMA Berta, und ich werde dich nie verlassen!
Sie geben sich die Hand. Soldaten, Feldesse.

BERTA Alma, ich muß dich was fragen. Sag mir grad, wie man das macht.

ALMA Was macht?

BERTA Daß man wen kennt.

ALMA Ich hab gleich einen.

BERTA So wie du möcht ich nicht sein. Einen Mann möcht ich schon kennen.

ALMA Du hast doch den Sohn von deiner Herrschaft.[2]

BERTA Das weiß ich nicht.

ALMA Das mußt doch kennen,[3] ob dir der was will.

BERTA *schupft die Achseln:* Ich hab mirs halt anders vorgestellt.

ALMA Aussuchen kannst du ihn dir nicht.

BERTA Aussuchen kann ich ihn mir schon. Einen Mann möcht ich kennen.

ALMA Sag gleich einen Pionier. Das ist doch nicht schwer.

BERTA Ich muß immer weglaufen, wenn mich einer ansieht.

ALMA Also, da gehst du zum Zuschaun, wenn sie die Brücke baun, eventunell. Da kommt so was von selber, eventunell.[4]

BERTA Was für eine Brücke?

ALMA Die neue über das Altwasser hinüber. Die Stadt liefert das Holz, und baun tun sie die Pioniere. Dafür muß die Stadt nicht zahlen.

BERTA Da geht den Großen wieder was hinaus.[5] *Beide ab.*

2

Zeck und Fabian haben ein Männergespräch. Beide kommen mit dem gleichen Hut daher, was am jungen Fabian seltsam aussieht.

ZECK In deinem Alter schmeiß ich so was schon lang.[6] In deinem Alter war ich ein Gelernter.

FABIAN Ich bring keine her.

ZECK Das hätte ich wieder nicht gesagt.

FABIAN Du, das muß man heraus haben. Die heutigen Mädel sind furchtbar.

ZECK Uns ist auch einmal angst gemacht worden.

FABIAN Wenn man ihnen was will, halten die einen zum Narren.

ZECK Das läßt man nicht auf sich sitzen.

FABIAN Da muß man denken und noch einmal denken.

ZECK Das ist falsch. Wie stellst dich denn an, Mann? Hast du keinen Schneid?

FABIAN Ich sags, wies ist, ich kann mich gar nicht mehr halten.
ZECK So bist du zum Haben.
FABIAN *prahlt:* Bei der nächsten, da pack ichs.
ZECK Augen auf, Ohren steif, und gleich muß es einem sein. Du wirst schon noch.[7]
FABIAN Da muß man an die Richtige hinkommen.
ZECK Die Berta – ist reif.
FABIAN Ich habe da meine Zweifel.
ZECK Da ist doch nichts dabei. Das Dienstmädchen hat man im Haus. Der kann man was mucken.[8] Das ist nicht wie bei einer Fremden. Mußt immer wissen, was du willst, Mann. Ist doch alles natürlich.
FABIAN Auf die Berta spitz ich mich schon lang.
ZECK Wenn das so ist, dann paß auf dich auf.
FABIAN Ich will aber nicht aufpassen. Ich stürze mich jetzt da hinein.
ZECK Das geht in den Graben. In der Liebe muß ein Mann kalt sein. Das muß er sich richten.
FABIAN Aber doch nicht gleich beim ersten Mal.
ZECK Hauptsächlich sind es in der Liebe die Fallen, wo die bewußten Fußangeln lauern. Für die Liebe ist das überhaupt charakteristisch.[9]
FABIAN Das lerne ich nie.
ZECK Die nackte Notwehr. Ich warne.
FABIAN Aber ist das nicht entsetzlich?
ZECK Du oder ich. Daran mußt du dich schon gewöhnen. Ein Mann verliert da seinen Kopf nicht.
FABIAN Du hast es mir nur schwerer gemacht.

3

Berta und Alma sitzen auf einer Glacisbank und singen ein Küchenmädchenlied.[10] *Sie halten sich an der Hand und gehn dabei mit dem Arm auf und nieder.*

Heinrich schlief bei seiner Neuvermählten,
einer reichen Erbin aus dem Rhein.

Böse Träume, die ihn immer quälten,
ließen ihn auch hier nicht schlafen ein.

ALMA Heut ist aber sehr ruhig da.
Singen.
Zwölf Uhr schlugs, es drang durch die Gardine
plötzlich eine kleine weiße Hand.
Was erblickt er, seine Wilhelmine,
die im Sterbekleide vor ihm stand.
Bebe nicht, sprach sie mit leiser Stimme,
Heinrich, mein Geliebter, bebe nicht.
Ich erscheine nicht vor dir im Grimme,
deiner neuen Liebe fluch ich nicht.
Pionier geht vorüber.

ALMA *steht auf:* Das war die Schwalbe.

BERTA Was machst du?

ALMA Es geht los.

BERTA Du kannst mir nicht wegaufen.

ALMA *dreht sich zu ihr um:* Ja, für dich heißt es springen.

BERTA *ruft ihr nach:* Alma!

ALMA Hilf dir selber. Von mir bekommst du keinen serviert.

BERTA So was von schlecht. *Berta geht in die entgegengesetzte Richtung zurück.*
Pioniere kommen vorbei. Münsterer, Roßkopf.

MÜNSTERER Bis wir um elf Uhr am Brückenkopf einpassieren, müssen wir eine gekannt[11] haben.

ROSSKOPF Das wird heute nichts mehr.

MÜNSTERER Da läuft noch eine. *Alma schlendert ihm in den Weg.*

ROSSKOPF Fräulein, warum gehen Sie so spät allein spazieren?

ALMA Heißen Sie Paul?

ROSSKOPF Warten Sie schon auf einen?

MÜNSTERER *reißt ihn weg:* Aber ich heiße Paul.

ALMA Weil ich da vorn zwei hab abfahren lassen, von denen hat der eine auch Paul geheißen.

MÜNSTERER Das sieht man Ihnen nicht an, daß Sie die Auswahl haben.[12]

ALMA Sind Sie nicht unverschämt, ich muß doch sehr bitten.

MÜNSTERER Gehn mer. Wir vertun da nicht unsere Zeit.

Beide ab.

ALMA *ruft ihm nach:* Leck mich am Arsch.

JÄGER *schiebt sein Fahrrad daher und läutet mit Absicht an seiner Klingel:* Fräulein, ich wäre heraufgekommen über den Berg, ohne daß ich absteigen muß. Aber wegen den Herren bin ich langsam gefahren. Nicht, daß Sie meinen, daß ich nicht radfahren kann.

ALMA Wenn Sie ein wenig warten, kommt ein schönes Fräulein vorbei, zu der sagen Sies wieder.

JÄGER Warum? Sie haben mich Ihnen schon so lang nachgehen lassen.

ALMA Das war eine andere.

JÄGER Das waren Sie.

ALMA Drehn Sie Ihre Laterne so, daß ich Ihr Gesicht sehe.

Er dreht sein Vorderrad.

Nicht wahr ist. Sie haben auch nicht die Stimme.

JÄGER Der, mit dem Sie sich verabredet haben, kommt doch nicht.

ALMA Der kommt.

JÄGER *fährt mit seinem Rad Kunstfiguren vor ihr:* Ohne Umfallen! So was muß man können. Das Rad gehorcht seinem Herrn.

ALMA *beeindruckt:* Wissen Sie, der, auf den ich warte, ist vielleicht schon vorbei, hat auch ganz anders ausgeschaut. Ich kann mich an den Mann schon gar nicht erinnern.

JÄGER Der Ärgste ist der, der nicht nachgibt.[13]

ALMA Es heißt so.

JÄGER Zigarette? *Er gibt ihr Feuer.* Ich habe auch schon öfter einem Mädel gefallen.

ALMA Ich sage ja nicht, daß Sie mir nicht gefallen.

JÄGER *klopft auf sein Rad:* Hopp !

Alma steigt hinter ihm auf sein Rad und fährt im Stehn mit ihm ab.

FELDWEBEL *kommt allein und geht allein. Ein Mädchen begegnet ihm, er will sich heranmachen, aber sie wendet sich von ihm ab:* Die Stadt ist nicht freundlich. *Ab.*

4

Korl und Berta treten auf.

BERTA Ich bin auch schon verraten worden.

KORL *zeigt nach vorn auf eine Bank:* Das ist unsere. Auf die haben wir gewartet.

BERTA Die gehört nicht daher. Die hat wer vertragen.

KORL Schaun wir, ob sie sauber ist. *Er leuchtet die Bank mit einem Zündholz ab.*

BERTA Ich hab mein helles Gewand an.

KORL Ich leg mein Taschentuch hin.

BERTA *setzt sich auf sein Taschentuch:* Gell, weils keine Lehne hat, drum ist nicht besetzt.

KORL Das ist doch nicht, wie wenn ich nicht dabei bin. Wofür habe ich meinen Arm? *Er legt seinen Arm um sie, sie lehnt sich daran zurück.*

BERTA *schmachtend:* Jetzt hab ich eine Lehne und weiß nicht, wie sie heißt.

KORL Korl.

BERTA *schmachtend:* Korl heißt sie.

KORL Da bist schon öfter gesessen.[14]

BERTA So nicht.

KORL Das machst dem nächsten weis, aber mir nicht.
Sie schweigt, er schubst sie ein wenig.
Sei kein Roß. So, dann bist du auch schon verraten worden.
Sie schweigt noch immer.
Auf einmal weißt du nichts.

BERTA Ich muß denken, wie ich es sage. Von keinem Herrn rede ich nicht. Von einem Mädel rede ich.

KORL *ungläubig:* Wie soll sie heißen?

BERTA Alma! *Trumpft auf.* Weil sie mich verraten hat. Zuerst hat sie mich brauchen können, und dann schaut sie mich nicht mehr an.

KORL Mir kannst du noch viel vorsagen. Das geht auch, ohne daß was wahr ist. Da muß nichts wahr sein.

BERTA Ist aber wahr.

KORL *schubst sie wieder:* Die Mädel meinen immer, sie können einen gleich zu was haben.[15]

BERTA Ich bin nicht wie die anderen.

KORL Meinst du! Ich bin wie alle anderen.

BERTA Das glaube ich nicht.

KORL Wie soll ich denn sein?

BERTA Weiß ich nicht.

KORL Dann kann ich mich aber nicht danach richten. Ich täte es schon nicht.

BERTA Magst ein Zündholz aufzünden, daß ich dich sehe?

Er leuchtet mit einem Zündholz sein Gesicht an und läßt das Holz ziemlich lang abbrennen.

Daß du dich fein nicht am Finger brennst.

KORL Da ist was dabei. Weil das ein Pionier nimmer spürt.

Er hält ihr seine Hand hin. Lang her auf die Haut, was das für eine ist.

BERTA *fühlt mit der Hand:* Schön.

KORL Hast einen Namen auch?

BERTA Eine Berta bin ich worden.

KORL Berta.

BERTA *eifersüchtig:* Hast schon eine gehabt, eine Berta?

KORL *spöttisch:* Pfeilgrad nein.

BERTA Das wär mir nicht recht, wenn du schon eine gehabt hast.

KORL Das täte ich dir nicht sagen. Muß man denn immer alles wissen vom andern?

BERTA Ja, mach dich schlecht.

KORL Da mach ich mich nicht lang schlecht. Weil ich sage, dir kann das gleich sein.

BERTA Mir ist es aber nicht gleich.

KORL *harter Griff:* Sag, daß es dir gleich ist!

BERTA Du tust mir weh.

KORL Das weiß ich.

BERTA Ich steh auf.

KORL Wenn ich dich laß! *Er versucht handgreiflich zu werden.*

BERTA Nicht! Das tut man nicht.

KORL Warum nicht? Das gwöhnst.[16]

BERTA *wird energisch:* Ich habe es nie getan. Ich geh sonst nicht mit die Herren. Wenn ein Herr so ist, zeige ich ihn an.

KORL *läßt ab:* Du kommst schon noch.

BERTA *ist aufgestanden:* Auf die Bank setze ich mich nicht noch einmal hin.

KORL Dann auf die nächste.

BERTA Korl, ich setz mich nicht hin.

KORL Warum, meinst, bin ich dann mit dir gangen?

BERTA *wie eine Beschwörung':*[17] Ich hör nicht hin, was er sagt, weil er das nicht so meint.

KORL Von dir laß ich mich lang verklären mit der Verklärung.

BERTA Gar nicht. Ich werde dich schon nicht verklären. Dich muß man schimpfen.

KORL Entweder es geht was zusammen oder es geht nichts zusammen. Wenn ein Mädel nicht zieht, tu ich nicht lang um. *Er geht fort, sie steht bestürzt und läuft ihm dann nach.*

BERTA Geh nicht weg. Denn ich will nicht mit dir streiten, und ich weiß nicht, wie das gekommen ist.

KORL Was ist jetzt? Stellen wir uns her, oder stellen wir uns nicht her?

BERTA *in die Enge getrieben:* Das kann ich auf einmal nicht sagen.

KORL Aber ich will dir auf einmal sagen, daß du nicht viel Zeit hast. Morgen kennen wir uns nicht mehr. *Ab.*

BERTA *ruft ihm nach:* Korl!

2. Bild
Haushalt Unertl

Eine Altane, mit Wâsche vollgehängt. Unertl in einem Schaukelstuhl. Fabian sieht in die Hausdächer.

UNERTL *ruft:* Berta! Bedienung! – Ich schrei mir den Hals aus.[18] Wo steckt dieses Weibsstück?

FABIAN Sie hat Ausgang genommen.

UNERTL Der gebe ich schon einen Ausgang. Ausgang hat das Weibsstück am Sonntag, nicht unter der Woch.

FABIAN Sie hat gesagt, sie muß an die Luft.

UNERTL Sie soll die Wäsche abnehmen auf der Altane, dann hat sie ihre Luft. Der Haushalt ist die gesündeste Abwechslung. Das ist statistisch erwiesen.

FABIAN Sie will eben auch einmal eine Ansprache haben.

UNERTL Was braucht die eine Ansprache? Die soll im Kopf haben, was von der Herrschaft verlangt wird. Alles andere ist Luxus.

FABIAN Wenn die Soldaten herumschwirren, leidet es sie nimmer daheim. Sie hat Angst, daß sie was versäumt.

UNERTL Die kommt noch früh genug zu ihrem ledigen Balg.

FABIAN Sie ist doch auch ein Mensch.

UNERTL Du mußt sie noch verteidigen.

FABIAN Die Berta ist anständig. Die Berta wirft sich nicht weg.

UNERTL Du kannst doch einen bei ihr markieren, wenn sie einen braucht. Dann wird sie nicht eigenmächtig. Niemand weiß was.

FABIAN Sie hat ihren Kopf auf.[19]

UNERTL Den muß man ihr austreiben. Das darf mir nicht einreißen, daß sich die ihren Feierabend macht und mich laßt sie hängen. Ich soll wohl noch in den Keller, wenn ich ein Bier will? Wofür hat man die Person?

FABIAN Sie arbeitet schon den ganzen Tag.

UNERTL Sie kann aus mir keinen Zimmerherrn machen, auf den man nicht aufpaßt. Bei mir daheim will ich mich pelzen. Ich will meine vier Wänd, wo mich niemand anscheißt. Meine Gewohnheiten leg ich nicht ab.

FABIAN Warum heiratest du nicht mehr?

UNERTL Ich bin ja noch bei Verstand.
FABIAN Von mir aus kannst du schon heiraten. Das macht mich noch nicht arm.
UNERTL So. Kann ich? Eine Frau springt noch lang nicht. Die Frau hat man am Hals. Bin ich verheiratet, dann hat die Mark fünfzig Pfennig.
FABIAN Du wirst schon nicht verhungern.
UNERTL Einmal die Pfoten verbrannt, und man langt nicht so leicht hin.[20]
FABIAN Auf die Mutter laß ich nichts kommen.
UNERTL Dich hat sie lieber gehabt wie mich.
FABIAN Das ist noch lang kein Verbrechen.
UNERTL Ich heirate eine Frau, die mich ersetzt in meinem Geschäft, und sonst nicht.
FABIAN Das muß es doch geben.
UNERTL Ich stell meine Ansprüch. Ich nimm keine, die nicht von hier fort war, hier war ich selber. Das muß schon eine sein, die mich hebt. Aber die nimmt mich wieder nicht. Warum weiß ich auch nicht.
FABIAN Einen wie dich wird keine nehmen, so wie du dastehst!
UNERTL Ich denke da nur an meine dritte gute Verkäuferin, die mir einen Korb gab. Die hat hinausgeschmeckt, die war im Ausland, die kannte sich aus. Sie hat einen gehabt, der ihr aus Amerika schreibt. Mir wäre Amerika ja zu weit. Das konnte sie näher haben. Aber nein!
FABIAN Hast du ihr einen richtigen Antrag gemacht? Ich kann es nicht glauben.
UNERTL Heiraten Sie in die Branche hinein, habe ich gesagt, wir zwei bringen es weit, gewandt sind Sie und alles. Das war doch ein Vorschlag. Mit Ihnen, habe ich gesagt, würde es mir nämlich passen.
FABIAN Das höre ich zum ersten Mal.
UNERTL Wenn Sie bei mir sind, habe ich gesagt, passiert Ihnen nichts mehr. Oder gerade dann, hat sie gesagt. Einfach mir ins Gesicht. Oder gerade dann.
FABIAN Die Person hatte an dich keinen Glauben.
UNERTL Kanaille! – Man darf eben keine nehmen vom Laden. Die

Mädchen wissen von einem zuviel. Heißt es ein ganzes Leben lang, keine zieht.[21] Die halten sich für strapaziert.

FABIAN Als hätten sie es dann schlechter wie zuvor.

UNERTL Als Verkäuferin, hat sie gesagt, ist es ihr bei mir lieber, weil das nichts Endgültiges ist. Es ist nicht das Ende. Da hast du es. Ich bin soviel wie das Ende. Ich weiß gar nicht warum.

FABIAN Du denkst nur an dich selber.

UNERTL Das machen alle. Auch wenn sie es nicht raushängen lassen.

FABIAN Ja, aber man merkt es bei dir.

UNERTL Dann eben nicht, schöne Tante! Ich reiße mich nicht darum. Wenn es nach mir geht, ich warte drei Jahre zu, und bau mir inzwischen den Großhandel auf, dann bin ich zweiundvierzig, von mir aus. Dann bin ich noch lang kein überständiger Mann. Ich muß auch so keinen Zimmerherrn machen. Ich kann mir ein Dienstmädel halten, die soll nur springen.

3. Bild
Schwimmbad Männerturnverein

Umkleidekabinen. Ein Holzsteg.

ZECK *aus einer Kabine:* Ich höre zuwenig. Wie weit bist du mit deiner Bestimmten?

FABIAN *aus einer Kabine:* Ich frage dich auch nicht nach deiner Bestimmten.

ZECK Frechwerden steht dir noch nicht zu.[22]

FABIAN Nicht hetzen, Häuptling. *Fabian kommt heraus.* Und überhaupt. Ich bin auf dem Weg.

ZECK Dann bleib du nur nicht pappen am Weg.

FABIAN Ihr sollt mich nicht aufziehn. Ich kann mich nicht beklagen. Sie wird meine Flamme. Es paßt.

ZECK *kommt heraus. Skeptisch:* Woran merkst du es, Mann?

FABIAN Ich habe sichere Anzeichen dafür.

ZECK Unser Wüstling! Hehe!

FABIAN Ich kann verlangen, daß man mich ernst nimmt.

ZECK Nur wenn du uns ihren persönlichen Unterrock bringst.

Bibrich balanciert über den Steg, gerät mit dem Fuß in ein Loch und fällt.

BIBRICH Oh, ah! Lieblich!

FABIAN Es hat einen geschnappt.

ZECK Und das passiert einem Schreiner.

FABIAN Weiß du nicht, jedes einzelne Brett liegt auf der Lauer?

ZECK Kannst du Büffel nicht schaun, wo du trittst?

BIBRICH Euch hör ich zu gern. Nur so weiter. Oh, Oh!

FABIAN Ist der Fuß ab?

BIBRICH Laßt mich liegen, ich weiß nicht.

ZECK Die Englein werden es ihm schon noch singen.[23]

BIBRICH Ich garantiere für nichts.

ZECK Schau nicht hin, da liegt eine Leiche.

BIBRICH Gemein seid ihr. Ihr seid gemein.

ZECK Du reust mich überhaupt nicht. Wir zum Beispiel können umgehn mit einem kaputten Steg. Bei uns gibt er nach.

BIBRICH Der Steg wird euch schon noch schnappen.

FABIAN Lieber nicht.

ZECK Ich muß zugeben, das wäre der dritte gebrochene Fuß. Der ganze Sausteg ist eine Falle.

BIBRICH Der Verein bringt keinen Steg auf die Welt.

ZECK Der Verein hat hinten und vorn nichts.

BIBRICH Für uns hat er nichts. Das sind doch Hirschen. *Er hinkt nach vorn.*

ZECK Jeden Monat hebt der Trainer die Hand auf. Der Trainer geht immer noch vor dem Steg.

BIBRICH Der Verein soll sich rühren, verdammt. Ich stoße mir nicht meine Glieder zuschanden.

ZECK Der Verein hat schon viermal bei der Stadt einen Vorstoß gemacht für einen Steg. Auf dem Ohr hört die Stadt nicht.[24]

FABIAN Der Schwimmer, der ist der Dumme.

BIBRICH Man kann nicht einmal ein Schauspringen organisieren, wenn man nichts hat, worauf man den Zuschauer stellt. Ohne Schauspringen kein Eintritt.

ZECK Ohne Eintritt keine Marie.

FABIAN Man kann es sich nicht malen.

ZECK Das sind Kasematten, mein Lieber. Das war ein Verteidigungsgraben. Für Sport und Schau ist das nicht gemacht. Wir haben genommen, was da war.

FABIAN Als Schwimmer gäbe es uns nicht.

BIBRICH Wer ist schon der Schwimmer? Der Schwimmer ist im Verein doch nur der kranke Mann.

ZECK Nur die Fußballer ziehen das Geld an.

FABIAN Uns geben die Fußballer nichts.

BIBRICH Der Schwimmer kommt nie zu seinem Steg.

ZECK Wir haben überhaupt keine Aussicht.

BIBRICH Können nicht ausweichen. Die Donau ist ja kein Wasser.

FABIAN Zuviel Wirbel. Die Donau täuscht.

ZECK Die Donau fälscht jedes Ergebnis. Wir sind schon angewiesen auf unseren langsamen Graben.

BIBRICH Die Stadt soll sich die Schwimmer am Hut stecken.[25] Wir streiken.

ZECK Seit wann tut das der Stadt weh? Die vermissen uns gar nicht.

FABIAN Wir kommen nur aus der Form.

ZECK Die können noch anders. Die hängen uns noch ein Schild hin: Zutritt verboten. Wir sitzen da und haben nicht einmal einen baufälligen Steg und haben kein Bad.

BIBRICH Dann wären wir die Geprellten.

ZECK Was machen wir dann?

FABIAN Es wäre das Ende. Und nur, weil man den Schwimmer bescheißt.

BIBRICH Dann muß ich schon sagen, wir basteln uns einen Steg. Ich als Schreiner würde euch das nämlich zeigen.

ZECK Holz müßte man haben.

BIBRICH Überstunden machen.

ZECK Werkzeug kann man sich leihn.

BIBRICH Man müßte schon schuften. Aber wir haben kein Holz.

ZECK Ich weiß ein Holz, wer sich traut. Das Holz ist nicht gesetzlich.

FABIAN *großzügig:* Pfeif!

BIBRICH Man muß sich schon traun.

FABIAN Unter uns ist kein Schuft, da kommt es nicht auf.

ZECK Am gleichen Altwasser, keine Viertelstunde von hier, liegt ein Holz. Man muß es bloß durchziehn.

BIBRICH Der Pionier ist im Land!

ZECK Der Pionier baut eine Brücke. Die Brücke gibt er der Stadt. Von der Stadt bekommt er das Holz.

FABIAN Die Stadt sitzt auf dem Holz.

BIBRICH Die Stadt gibt dem Schwimmer das Holz nicht.

ZECK Der Schwimmer nimmt sich das Holz.

BIBRICH Das ist nicht mehr wie gerecht.

FABIAN Weil ihm das Wasser bis an den Hals steht.

BIBRICH Der Verein darf nichts wissen.

ZECK Der Steg ist nicht einmal ein kleiner Wald. Einen kleinen Wald muß die Stadt sich eben leisten.

4. Bild
Bierzelt

Bürger und Pioniere. Berta, Fabian und Feldwebel an einem Tisch. Korl kommt herein und vermeidet es, den Feldwebel zu grüßen. Er setzt sich an einen anderen Tisch. Aber der Feldwebel hat es bemerkt. Außen in der Nähe des Bierzelts eine Rotunde mit der Aufschrift Pissoir.

FELDWEBEL Der Kerl ist auf dem Auge blind,[26] aber ich will nicht so sein. Ich könnte es ihm nämlich geben.

FABIAN Sie sind kein Spielverderber.

FELDWEBEL *heuchelt:* Nie. Wie sagt man, ein Maß?[27]

FABIAN Eine Maß für das Militär! Nämlich weiblich. Das Ding ist bei uns weiblich.

FELDWEBEL Haben Sie den ganzen Tag die Spendierhosen an?

FABIAN Es geht. Zahlt alles mein Vater. Sie brauchen sich nichts dabei denken.

FELDWEBEL Denken ist vom Übel. Ich revanchiere mich noch.

FABIAN Ich komme darauf zurück.

FELDWEBEL Hier stimmt was nicht. Warum sitzt die taufrische Begleitung[28] auf dem verkehrten Platz? Das Militär hat nichts davon. Sie müßte zwischen uns sitzen.

FABIAN Das Mädchen sitzt schon ganz richtig.

FELDWEBEL Das glaube ich nicht, weil sie stumm ist. Haben Sie sich gestritten?

FABIAN Wir streiten nie.

FELDWEBEL Ich zum Beispiel wäre nicht abgeneigt, das Mädchen zu kennen. Sie wäre weniger stumm.

FABIAN Vielleicht wäre sie abgeneigt.

FELDWEBEL Ich fürchte eher, der Zorn geht auf Sie.[29]

FABIAN Auf mich? Pah! Sie frißt mir aus der Hand.

FELDWEBEL Es sieht mir nicht danach aus.

FABIAN Sie blamieren mich, Berta. Wollen Sie nichts zu sich nehmen?

BERTA Nichts. Danke.

FABIAN Warum sind Sie dann mit mir hergegangen, Berta, wenn

Sie mich nicht anschaun?

BERTA Ich kann nicht die ganze Zeit auf Sie hinschaun. Das fällt auf.

FABIAN Das soll auffallen.

BERTA Da muß einer danach sein auch.

FELDWEBEL Das würde ich mir nicht im Schlaf nachsagen lassen.[30] Da muß ich schon hetzen.

FABIAN Sie, ich täte nicht auslassen bei einem Mann, wie ich bin.

BERTA Wer sind Sie schon?

FABIAN Berta, das war eine Kränkung.

Korl pfeift ihr.

Was soll das heißen?

Berta will aufstehn, Fabian hält sie zurück. Korl pfeift noch mal und macht ihr ein Zeichen herüberzukommen.

FELDWEBEL Frechheit! Vom Tisch weg!

FABIAN Seit wann kennen Sie den? Jetzt gehn Sie grad nicht hin, weil er meint, es muß sein.

BERTA Da muß ich schon hin. *Sie macht sich los und geht zu Korl an den anderen Tisch.*

FELDWEBEL Aber das kann sie nicht machen. Holen Sie Ihre Braut doch zurück.

FABIAN Sie ist nicht meine Braut.

FELDWEBEL Wir können schaun wie die Affen.

FABIAN Wir können uns noch immer besaufen.

Sie stürzen Bier hinunter.

FELDWEBEL Wenn einer über mich lachen kann, das habe ich nicht gern.

FABIAN Über Sie lacht er ja gar nicht.

FELDWEBEL Ich kenne den Mann. Der Kerl ist zu schnell am Ball und schnappt seinem Vorgesetzten die Jungfrauen weg. Das war schon so in Küstrin, und jetzt hängt er mir zum Hals heraus, dem Mann sitze ich auf.

Musik: Stolz weht die Flagge Schwarz-Weiß-Rot.[31]

BERTA Gell, die mit den Achseln sind andere, wie du bist.

KORL Das sind meine Vorgesetzten.

BERTA Wie heißt man das, wenn sie deine Vorgesetzten sind?

KORL Das heißt man Unteroffizier, und das heißt man Feldwebel.

Sprich Uffz und Feld. Oberfeld und die ganze Leier hinauf bis zum General, der ein alter Hirsch ist und der einen einschnauft wie Luft.

BERTA Sprich Uffz und Feld! *Sie lacht.* Aber ein Pionier ist viel schöner wie ein Feldwebel.

KORL Weil er der Jüngere ist. Beim Militär ist das ein Nachteil. Unsere Vorgesetzten sagen, ein Pionier, wenn er nicht zehn Jahre lang einer gewesen ist, ist überhaupt noch kein Pionier

PIONIER Überhaupt noch keiner.

BERTA So schwer ist das.

KORL Ein Pionier muß viel mehr können wie seine Vorgesetzten, aber er darf es nicht merken. *Er schließt Berta in die Arme.*

FABIAN Wie heißt du, Feldwebel?

FELDWEBEL Willi.

FABIAN Ich werde Willi zu dir sagen. Ich zahle dir auch noch eine Maß.

FELDWEBEL Weiblich. Man merkt, daß du einer von den Großen bist.

FABIAN Ich bin erst am schwachen Anfang, aber ich bin schon drin in der Bahn.

FELDWEBEL Kunststück!

FABIAN Ich komme nach meinem Alten. Ich wachse hinein in die Firma. Einmal bin ich der Chef.

FELDWEBEL Bei dir kann nichts schiefgehn.

FABIAN Die Mädchen müßten einem nachlaufen, hätten sie einen Verstand.

FELDWEBEL Die Mädchen laufen immer den Falschen nach.

FABIAN *trinkt:* In so einen Maßkrug geht allerhand hinein.

FELDWEBEL Darf schon was hineingehn. Ich schwemme meinen ganzen persönlichen Verdruß hinunter.

FABIAN Saufbruder, du auch?

FELDWEBEL Du ahnst es nicht.

FABIAN Saufbruder, sauf!

Verbrüderung. Kriegerisch.

Auf der Welt!

FELDWEBEL *kriegerisch:* Auf der Welt![32]

FABIAN Die Letzten werden die Ersten sein.[33]

FELDWEBEL Ich kann das von hier aus nicht sehn. Weil wir alle Arschlöcher sind, jawoll, und weil der Druck nach unten geht. Und weil sie mich gestaucht haben wegen dem Holz, jawoll.

FABIAN Hast du Holz gesagt?

FELDWEBEL Ich sage Holz, aber es ist schon eine halbe Brücke.

FABIAN Sprich nicht weiter. Ich habe einen Verdacht.

FELDWEBEL Ich habe immer einen Verdacht, das muß ich. Mir macht es nicht so leicht einer recht. Das ist meine verdammte Pflicht. Ich muß sein wie eine Geißel.

FABIAN Davon verstehe ich nichts.

FELDWEBEL Und darum hast du es gut.

FABIAN Ich habe einen schwarzen Verdacht, mein kleiner Finger will mir was sagen.[34]

FELDWEBEL Ich bin nicht dein kleiner Finger.

FABIAN Sei nicht so.

FELDWEBEL Ich könnte platzen. Du siehst einen Mann vor dir, dem sie die halbe Brücke gestohlen haben, weil er das Kommando vom Bautrupp hat. Einem Zivil kann ich das gar nicht näher erklären. Du siehst einen geschlagenen Mann.

FABIAN Ich habe was läuten gehört von verschwundenem Holz,[35] und ich habe es geglaubt, so bin ich.

FELDWEBEL Wie vom Erdboden verschluckt, und dabei war es schwer.

FABIAN Die Menschen sind ja so schlecht.

FELDWEBEL Und das ist noch gar nichts. Ich habe einen Rüffel einstecken müssen, den man sich nur unter Zwang ausdenken kann. Du siehst einen Mann vor dir, der hat nichts mehr zu lachen.

FABIAN Wird denn so ein Haufen Holz nicht bewacht?

FELDWEBEL Wer läßt bewachen, was man nicht einmal schleppen kann? Es bewacht sich von selber.

FABIAN Ein Mann allein bringt es nicht fertig.

FELDWEBEL Das war schon mehr eine Bande. Ich muß Leute haben in meinem eigenen Trupp, die haben es vom Fleck weg verkauft.

FABIAN Organisiert.

FELDWEBEL Geraubt. Den Kopf reiße ich ihnen herunter.[36]

FABIAN Den Kopf reißt du ihnen herunter.

FELDWEBEL Erst muß ich sie haben. Und ich werde sie haben. Es sticht mir in die Nase, wer es sein muß,[37] aber ich kann nicht an sie heran. Ich kann nicht sagen von Mann zu Mann, Hund, das warst nur du. Ich fühle meine Ohnmacht, auch wenn ich sie zudecke mit Schikanen.

FABIAN Wem willst du es anhängen?

FELDWEBEL Wenn man sucht, findet man immer was.

FABIAN Dann baust du nicht weiter, Boß?

FELDWEBEL Ich baue immer weiter, verdammt!

FABIAN Wer ersetzt dir das fehlende Holz?

FELDWEBEL Die Stadt. Wer sonst?

FABIAN Die Stadt wird euch was blasen.[38]

FELDWEBEL Wir werden der Stadt was blasen. Will sie ihre Scheißbrücke über den Graben, oder will sie ihre Scheißbrücke nicht?

FABIAN Vielleicht hat sie sie nicht verdient. Vielleicht hätten sie ganz andere verdient.

FELDWEBEL Immerzu wird die Stadt es ersetzen. Aber die Stadt sieht es nicht gern. In der Stadt sind sie nicht freundlich.

FABIAN Das geht alles seinen Gang.

FELDWEBEL Mich kostet es meine Beförderung, das ist der Gang. Man hat mich zum Schuldigen gemacht, das ist der Gang. In solchen Fällen wird der General ein Stier, und der Major wird ein Stier, und der Hauptmann wird ein noch größerer Stier. Je mehr nach unten, desto reißender der Zorn, und desto mehr wirkt es sich aus. Der Druck geht nach unten.

FABIAN Was machst du damit?

FELDWEBEL Ich gebe ihn weiter, den Druck.

FABIAN Siehst du!

FELDWEBEL Und ich werde einen Schuldigen finden, und wenn er nicht schuldig ist, dann mache ich ihn schuldig.

FABIAN *hält ihm Bier hin:* Wir zwei Hübschen können noch immer ersaufen.

FELDWEBEL Es hilft zuwenig. Der Wurm ist schon drin.

FABIAN Der Gefreite, der jetzt mit dem Mädel ist, was ist das für einer privat?

FELDWEBEL Mit einem Pionier rede ich nicht.

FABIAN Aber das Mädel hat dir entsprochen?

FELDWEBEL Wenn das Mädel einen Pionier vorzieht, kann ich mich nicht damit befassen.

FABIAN Muß der auch im Lokal strammstehn vor dir? Das möchte ich sehn.

FELDWEBEL Das wirst du auch sehn. Denn das ist auch so einer, der klaut und der sich alles unter den Nagel reißt, und nichts ist ihm heilig.

Musik: Holzhackergesellen.[39] *Korl verläßt mit Berta das Zelt. Er hat wieder nicht gegrüßt. Der Feldwebel steht drohend auf.*

KORL *draußen:* Ich muß Bier abgießen, Mädchen.[40]

Er verschwindet in der Rotunde, Berta wartet am Zelt. Der Feldwebel sucht gleichfalls die Rotunde auf. Er kommt dicht hinter Korl wieder heraus.

FELDWEBEL Der Gefreite, he! Seit wann ist Ihr Vorgesetzter für Sie Luft?

KORL Ich habe den Feld nicht gesehen, zu Befehl.

FELDWEBEL Sie grüßen mir genauso wie ein Rekrut, ich lasse Ihnen nichts durchgehn, und wir holen das jetzt hübsch nach. Ich zum Beispiel will von Ihnen einen Gruß haben, der sich gewaschen hat.[41] Achtung – *Korl muß grüßen.*

Das ist für mich nicht einmal der schwache Entwurf zu einem vorschriftsmäßigen Gruß. Das Ganze nochmal. Achtung – Achtung – *Korl muß grüßen.* So geht das nicht, Mann. Sie müssen sich hineinwerfen mit allem Drum und Dran. Das muß nur so spritzen. Das muß überspringen aus Ihrem morschen Gestell. Ihnen mache ich Feuer unter dem Arsch. Und zurück! Zack und zack und zack -

Er sprengt Korl auf und ab, wobei dieser grüßen muß. Während der Schikane ist ein Mann aufs Pissoir gegangen und sieht sich die Schikane an. Aus dem Zelt glotzen Bürger heraus, Fabian. Einer singt anscheinend unmotiviert:

Die Fliegen, die Fliegen, die kann man halt nicht kriegen.[42]

BERTA Ich würde mich schämen. So lassen Sie den Mann doch aus.

FELDWEBEL Das Mädel wird Ihnen nichts helfen. Die soll sehn, wer den kürzeren zieht.

BERTA Sie sind ein schlechter Mensch.
FELDWEBEL Lassen Sie nur das Mädchen.
KORL Das ist mein Mädchen, zu Befehl.
FELDWEBEL Das sind keine Mädchen für die Mannschaft. Das sind Mädchen für die Dienstgrade, verstanden.
KORL Ich habe Ausgang, zu Befehl. Wenn ich Ausgang habe, lache ich mir eine an, das macht sich von selber.
FELDWEBEL Das macht sich nicht von selber. Zum Beispiel kostet es mehr Geld, als Sie haben. Ich persönlich halte das für verdächtig.[43]
KORL Bei mir ist das anders. Ich bekomme so was umsonst.
FELDWEBEL *hat daran schwer zu schlucken,*[44] *er rächt sich:* Hinlegen! Robben!
Korl muß sich hinlegen und robben.
BÜRGER Warum denn? Unerhört! Schikane!
FELDWEBEL Die Stadt ist nicht freundlich.
ALMA *schlendert heran:* Warum so stürmisch, mein Freund?
FELDWEBEL Ich bin nicht gleich so. Aber wenn ich in Trab gesetzt werde, stampfe ich einen zusammen.
ALMA Aber doch nicht vor meinen sehenden Augen.
FELDWEBEL *wittert seine Chance:* Aufstehn! Abtreten!
Korl steht auf. Der Feldwebel gibt ihm ein Zeichen, daß er entlassen ist. Korl setzt sich mit Berta auf eine alleinstehende Bank. Fabian geht ins Zelt und säuft.
ALMA *zu Berta:* Dein Kavalier kann sich bei mir bedanken.
Das Publikum verläuft sich.
ALMA *zum Feldwebel:* Na, wie gehts?
FELDWEBEL Nicht mehr so oft wie früher.[45]
ALMA Das bilden Sie sich bloß ein.
FELDWEBEL Gut schaun Sie aus, Fräulein. Und an mir ist eine vorbeigangen, die hat gesagt, ich bin nicht mehr zum Kennen. Ich kann mir selber nicht gut sein.
ALMA Das ist lebensgefährlich.
FELDWEBEL Sie sehen in zehn Jahren auch nicht mehr so aus.
ALMA Was geht das mich an? Dann habe ich eben jetzt meine große Zeit.
FELDWEBEL Wie alt sind Sie, Fräulein?

ALMA Zwanzig.

FELDWEBEL Das geht gerade noch. Wo wohnen Sie, Fräulein?

ALMA Am Unteren Graben.

FELDWEBEL Akkurat in meiner Nähe.

ALMA Ich bin sehr beliebt bei den Herren. Ich kenne die meisten. Ich bin überhaupt eine mondäne Frau.

FELDWEBEL Ich schrecke vor nichts zurück.

ALMA Aber ich fliege nicht auf einen jeden.

FELDWEBEL Warum so hart? Man kann doch eine Ausnahme machen. Ich habe eine Wut im Bauch. Fräulein, richten Sie mich wieder zusammen.

ALMA Nicht für deine schönen Augen, wenn du das meinst. Nämlich es gibt eine Kasse, mein Freund. Ich bin ohne Stellung.

FELDWEBEL Das hört er nicht gern.

ALMA Jung sind wir nur einmal.

FELDWEBEL Du machst es gnädig, was?

ALMA Und ich bleibe nichts schuldig.

FELDWEBEL Ich möchte sagen, das fehlt mir gerade. *Beide ab. Korl und Berta auf der Bank.*

KORL Tu dich nicht in mich verlieben, Kind.

BERTA Ich verliebe mich nicht.

KORL Das haben schon viele gesagt und haben sich doch in mich verliebt.

BERTA Gell, Korl, wir machen Spaß.

KORL Ich mache keinen Spaß. Mich muß man laufenlassen.

BERTA Du bist dumm. Dich will ich gerade. Dich habe ich mir ausgesucht von alle.

KORL Tu dich nicht in mich verlieben, sonst muß du leiden.

BERTA Ich will leiden.

KORL Du kennst mich nicht. Da kann ich bös sein, wenn eine gut zu mir ist. Die Frau wird von mir am Boden zerstört, verstehst. Da kenn ich keinen Bahnhof.

BERTA Du tusts halt gern.

KORL So kannst auch sagen. Ist doch bloß, daß man weiß, was es alles gibt.

BERTA Wenn ich aber in dich verliebt sein will.

KORL Das weiß ich, sonst bist du nicht ganz. Aber ich habe es dir gesagt.

BERTA Dir tut es auch gut, wenn der Mensch bei einem Menschen ist.

Pause.

Um gar nichts ist mir angst, als daß dir eine andere gefallt.

KORL Haben wir schon einen Krampf. Das kann ich nicht leiden. Red nicht, weils nichts wird. Mach dich ein wenig leicht.

Er fummelt ihr am Hals.

BERTA Die Sterne scheinen darauf. In deine Augen scheinen sie auch.[46]

KORL Paß auf und red nicht.

BERTA Korl, ist das schön?

KORL Jetzt höre ich auf, weil ich nicht mehr mag. So laß ich mich nicht ansingen.

BERTA *bedeckt sich:* Soll ich gehn?

KORL Ja, geh. Ich hab keinen Magen.

Berta bleibt.

Siehst, auf einmal kann ich nicht mehr mögen, so bin ich.

BERTA Du kannst es, aber dann ist dir was auskommen.

KORL Da ist mir nichts auskommen.

BERTA Es ist was mit dem Herzen, und du weißt es nicht.

Berta ab.

Pause.

5. Bild
Schwimmbad Männerturnverein

Zeck macht Freiübungen. Hinten im Wasser Schwimmgeräusche eines unsichtbaren Schwimmers.

BIBRICH *tritt auf:* Wen hast du im Wasser?

ZECK Er weiß nichts. Es ist bloß der Ratz. Er ist jeden Tag hier.

BIBRICH Ist der sicher?

ZECK Er ist der Jüngste und weiß, was ihm passiert im Verein. Auf den drückt man den Daumen.

BIBRICH Mit dem Holz muß es stimmen. Die Polizei schwirrt herum, ist schon draußen am Platz. Hier herunter kommen sie auch.

ZECK Man sieht es nicht ein. Es liegt eingekeilt an der Mauer im Schatten. Das Holz liegt ihnen zu tief.

BIBRICH Die Polizei müßte schon tauchen.

ZECK Bei uns taucht die Polizei nicht. Sie haben überhaupt keinen Anhalt.

POLIZIST *tritt auf:* Routineuntersuchung, meine Herren. Dieser Teil vom Gelände ist noch nicht geprüft.

ZECK Die Polizei wird uns doch nichts wollen.

POLIZIST Die Polizei kann keine Ausnahme machen.

ZECK *führt ihn herum:* Vorsicht am Steg. Da sind Bretter gelegt. Die werden erst noch genagelt.

POLIZIST *aufkeimender Verdacht:* Ich sehe.

ZECK Unser Schreiner ist nicht mobil.

Bibrich leidet.

POLIZIST *bückt sich:* Die Bretter sind alt. *Er schreibt.*

ZECK Von der Brauerei. Eine Spende.

POLIZIST *öffnet die Kabinentüren:* Was ist hinter den Türen?

ZECK Wir haben hier keine Damen.

POLIZIST Damen suchen wir nicht. Auf welchen Fluchtweg führt die Leiter?

Eine Leiter steht angelehnt neben den Kabinen.

ZECK Die Leiter geht nur auf den Wall.

POLIZIST Ich überzeuge mich, was am Wall ist. *Er steigt auf den*

Festungswall.

ZECK *hinterfotzig:* Es ist der höchste Punkt. Von oben sieht man herunter.

POLIZIST *überblickt das Gelände und steigt dann herunter:* Wall unterbrochen. Kein Fluchtweg. *Er schriebt in sein Buch.* Wie tief ist das Wasser?

ZECK Vier Meter vielleicht.

POLIZIST Informant bezeichnet die Wassertiefe mit vier Meter.

ZECK Kann mehr sein. Auf den Zentimeter habe ich es nicht gemessen. Sie müßten schon tauchen.

Bibrich leidet.

POLIZIST *schreibt:* Name des Informanten – wie heißen Sie?

ZECK Ludwig Zeck.

POLIZIST Geboren?

ZECK 1899 22. November hier.

POLIZIST Ständiger Wohnsitz?

ZECK Goldknopfgasse 7. Der Verein hat mich in der Kartei.

POLIZIST *beugt sich über das Wasser:* Man sieht nicht auf den Grund.

ZECK Hier sieht man nie auf den Grund. – Haben Sie schon einmal Schokolade unter Wasser gegessen?

POLIZIST Nein. Warum?

ZECK Schmeckt nach gar nichts. Im Mund sind lauter Scherben.

POLIZIST Interessant. Ich bin kein Taucher. *Grüßt und geht ab.*

ZECK Der Hirsch !

BIBRICH Der Schutzmann hat uns überhaupt nicht gefilzt. Das war Massel.

ZECK Das Holz soll uns schon bleiben.

BIBRICH Nach und nach holen wir es dann herauf.

ZECK Es wird uns doch im Wasser nicht faulen.

BIBRICH Das Holz ist präpariert.

ZECK Bis der Pionier fort ist, müssen wir es schon verstecken.

BIBRICH Ich bin ganz abgeschlagen. Mich hats geschlaucht.

ZECK Ich rolle mich auch. *Beide ab.*

6. Bild
Baustelle

Nacht kurz vor Passierschluß. Brückenbauskelett. Korl und Bunny nähern sich vorsichtig.

KORL Ich kann den Kerl mir nicht mehr aufsitzen lassen, und ich muß ihn mir aufsitzen lassen. Bunny, es macht mich krank.

BUNNY Ich mag ihn ja auch nicht, weil er ein Radfahrer ist.

KORL *klettert am Brückenskelett hoch:* Wenn einer kommt, unseren Vogelpfiff. *Er klettert höher und hantiert an der Verschraubung.*

BUNNY Du kannst nicht alle Schrauben lockern, spinnst du, das fällt auf. Du reitest dich in was hinein.

KORL Unterbrich mich nicht, Bunny.

BUNNY Es ist mindestens Sabotage. Weißt du, was darauf steht?

KORL Der Schinder soll merken, daß zwischen ihm und der Mannschaft Krieg ist. *Er ist wieder unten.*

BUNNY Wenn er filzt, kommt er doch nur auf dich.

KORL Er wird es nie wirklich wissen.

Bunny pfeift leise den Vogelpfiff.

Korl fährt herum und kann niemand entdecken. Was pfeifst du, Idiot?

BUNNY Ich mache dir Angst. Nichts wie weg. Er sitzt am langen Arm, du kommst immer zu kurz. Du bist das arme Schwein, sieh dich vor. Er wird Metzger spielen.

KORL So einer wäre schon lang weggeputzt von hinten in einem Krieg.

Sie schleichen sich weg.

7. Bild
Haushalt Unertl

Unertl, Berta, Fabian

UNERTL Es wundert mich, daß wir für Sie überhaupt existieren. Wir zahlen ja nur den Lohn. Wir sind die reinsten Waisenknaben gegen das Militär.[47]

BERTA Ich tu meine Arbeit. Der Herr kann sich nicht beklagen.

UNERTL Sie können nicht zählen. In der Woche war es das zweite Mal, daß das Fräulein ausgeflogen war, und man hatte das Fräulein nicht bei der Hand.

BERTA Ich hab noch jeden Abend abgespült nach dem Essen. Danach habe ich auch einmal frei.

UNERTL Die Arbeit darf nicht darunter leiden. *Greift aus Schikane nach einem Teller. Zu Fabian.* Schau dir den Rand an, der da sitzt.

BERTA Da sitzt kein Rand.

UNERTL Weil sie kein Wasser heiß werden läßt vor lauter Pressieren.

FABIAN Die Zeichnung hat das Geschirr von jeher gehabt.

BERTA Immer kommt man in einen falschen Verdacht.

UNERTL Der Mensch drückt sich aus durch seine Arbeit. Die Vorhänge könnten Sie schon lang herunternehmen zum Beispiel.

BERTA Ich habe vor vierzehn Tagen gewaschen. Die Vorhänge auch.

UNERTL Das muß blütenweiß sein, das muß nur so funkeln. Die Sauberkeit muß einem ins Gesicht springen, das heiße ich einen Haushalt führen. Das heiße ich einen Kampf mit den Bazillen.

BERTA Ich kann nicht jeden Tag das Bett überziehn.

UNERTL Bilden Sie sich nur nicht ein, bei mir können Sie mogeln. Wenn ich schon zahle, hole ich den Gegenwert aus einem Menschen heraus, ich wäre ja dumm.

BERTA Immer drücken Sie drauf, sprengen einen herum.

UNERTL In einem Haushalt gibt es noch immer was zu putzen, wenn man danach sucht und wenn man sich einkrallt. Das ist eine Aufgabe fürs ganze Leben.

BERTA Davon wird man gefressen.[48]

UNERTL Wenn Sie das nicht tun, sind Sie für mich eine ganz gewöhnliche Schlampe, keiner weint Ihnen nach.

BERTA Sie verlangen zuviel.

UNERTL Ich weiß schon, Sie sind geschützt.[49]

BERTA Sie müssen einen schon schnaufen lassen.

UNERTL Daß Sie mir nicht damit aufs Arbeitsamt laufen! Sie würden sich bloß blamieren. Sie wissen gar nicht, was man da sagt. Wenn Sie mir was anhängen, dann machen Sie sich auf was gefaßt.

BERTA Ich kann mich nie rühren, daß man was von mir merkt.

UNERTL Von Ihnen braucht man noch lang nichts merken.

BERTA Ich bin doch auch ein Mensch.

UNERTL Weil die junge Generation nicht mehr weiß, was man ihr so schon hineinsteckt und wie gut sie es hat.

BERTA Ich muß mir nicht alles gefallen lassen.

UNERTL Eine Minute nicht den Aufpasser gemacht, und schon ist sie draußen beim Tempel.[50] Und dann kommt sie daher mit dem Verzug.

BERTA Mit welchem Verzug.

UNERTL Ich kann einen Lebenswandel verlangen, wenn ich wen aufnehme in meinem Haus. Da könnten Sie mir ja eine galante Krankheit daherbringen. Sie sind doch nicht läufig.

FABIAN Jetzt mach aber langsam.

UNERTL Halte dich du da heraus. – Man muß es Ihnen schon sagen, wenn Sie nicht wissen, was einem vertrauensseligen Mädchen passiert.

BERTA So schlecht ist keiner. Davor braucht einem nicht angst sein.

UNERTL Sie sind ja erst siebzehn. Sie wissen es besser.

BERTA Alles machen Sie einem schlecht. Der Herr soll mich weglassen. Ich kann den Herrn nicht mehr sehn.

UNERTL Sind Sie nicht vorlaut. Sie hat ein Maul wie ein Schwert.[51]

BERTA Der Herr hat alles an mir zum Aussetzen.

UNERTL Sie werden es brauchen.

BERTA Wenn ich auf den Gang gehe, schickt mir der Herr den Sohn nach.

UNERTL Kommst du überhaupt weiter damit? Ich merke nichts.

FABIAN Das kann man nicht vor dem Mädchen besprechen.

UNERTL Wir bringen das nämlich auch noch zusammen. So weit brauchen Sie dafür gar nicht laufen.

BERTA Ich muß schon bitten.

UNERTL Ihr Soldat hat das nicht gepachtet. Wo faßt er Ihnen denn hin?

FABIAN Du bist ordinär.

UNERTL Ich nehme mir da kein Blatt vor den Mund. Ich bin bloß ehrlich. Soll sie halt einmal nachgeben. Dann weht gleich ein anderer Wind.[52]

BERTA Weiß der Herr nicht, daß er sich damit was vergibt?

UNERTL Das geht auch anders herum. Sie bringe ich noch auf die Knie, Sie Person. Ich werde Sie schon sekkieren. Sie sollen spüren, daß man Sie in der Gewalt hat.

BERTA Wenn der Herr mich zwingen will, zwingen lasse ich mich nicht. *Ab.*

UNERTL Auf einmal zeigt sie die Krallen.

FABIAN So geht das nicht. Das mache ich schon selber.

UNERTL Mit der hast du es schon verpaßt. Die zieht nicht.

FABIAN Das Mädel kennt sich nicht mehr aus, wofür man sie hat.

UNERTL Laß sie laufen. Gibt doch andere genug.

FABIAN Ich bin dir nicht mehr gut.

8. Bild
Baustelle Brückenbau

Frühnebel. Pioniere arbeiten.

FELDWEBEL Ihr Lahmärsche arbeitet in Tagesschichten von sechs Stunden, das ist zu wenig. Ab heute Nachtschicht, sonst kommen wir nie durch. Hier wird geschuftet, das war nicht in den Wind gesprochen, ich spreche nie in den Wind. Steht nicht herum wie die Säcke. Ihr wollt doch vortäuschen, daß ihr was tut. Los, los, ein bißchen lebhaft da drüben, ich prüfe es nach. Meine Augen sind scharf, ich picke jeden einzeln heraus. *Eifrigeres Arbeitsgeräusch. Er klettert am Brückenskelett hinauf an seinen gewohnten Platz, wo er die bessere Übersicht hat, und stürzt ab. Es herrscht reine Schadenfreude. Ein paar Liebedienerische rennen zu ihm hinaus.*

ROSSKOPF Jetzt liegt die Sau im Schlamm.

BUNNY Wetten, daß er sich verzieht!

ROSSKOPF Da kennst du den Schinder schlecht.

BUNNY Ist er hoch?

KORL Und wie! Keine Müdigkeit vorschützen, jetzt sind wir dran.[53]

Sie arbeiten eifrig.

FELDWEBEL *zerschundene Uniform:* Mit mir nicht, du Hundling, ich prüfe das nach. Weitermachen! *Der Feldwebel klettert am Bauskelett hoch und untersucht die Ursache. Er steigt herunter, Schrauben in der Hand.* Ich betrachte das als persönlichen Angriff. Gruppe fünf antreten!

Gruppe fünf tritt an, darunter Korl, Münsterer, Roßkopf, Bunny.

Welcher gemeine Verbrecher hat die Schweinerei mit der Verschraubung gemacht? Sicherungsschrauben offen, das ist Sabotage. Das ist strafbar im höchsten Grad. Das ist ein heimtückischer Anschlag. Ich erwarte sofortige Meldung vom Schuldigen, wer war's? Wir kommen ihm ja doch darauf.

Schweigen.

Ich warne. Wenn keiner sich meldet, hat die ganze Gruppe zu

büßen. Die Schrauben waren gestern intakt. Wer hat gestern abend Ausgang gehabt? Er soll sich melden, ich prüfe das nach. *Er mustert Gruppe fünf* . Da haben wir unsere Hübschen beisammen. Die Hübschen werden wir filzen. *Dreht sich um.* Das ist kein Anlaß für Faulheit. Weitermachen dahinten! *Er geht kurz nach hinten.*

KORL Wer war es eigentlich? Will es keiner gewesen sein? Hat er Angst, wir stoßen ihm den Balken in seinen Rücken hinein?

ROSSKOPF Wir wissen es nicht. Von uns war es keiner.

MÜNSTERER Wir haben am Tag genug Gegend.[54]

ROSSKOPF Auf die Baustelle kommen wir überhaupt nicht heraus. Sie steht uns bis hier.[55]

MÜNSTERER Aber ich kann mir schon denken, wer ihn gern wegputzt.

KORL Denken nützt gar nichts. Wissen!

MÜNSTERER Einer spielt verrückt. Alle anderen fressen es aus.

KORL Mensch, wovon rede ich die ganze Zeit?

Sie messen sich unfreundlich.

FELDWEBEL *kommt zurück:* Ich erwarte, daß jeder Hundesohn den Beweis erbringt, wo er gestern seine freien Stunden verbracht hat. Wer den Zeugen nicht beibringt, der für ihn gutsteht, der kann sich freun. Wer keinen Zeugen hat, der sagt es mir besser gleich. Unser Mann ist noch nicht weich. Aber ich mache ihn weich. Und das gilt für die ganze Gruppe, weil sie den Schuldigen deckt. Gruppe fünf antreten zum Strafexerzieren. An den Balken heran!

Sie stellen sich hintereinander am Balken auf, der Kleinste nach hinten.

Den Balken – hebt. Mit dem Balken – marsch, marsch. Zurück, marsch, marsch. Knie beugt – streckt – Laufschritt, marsch, marsch! Ich habe Laufschritt gesagt. Abteilung kehrt. Laufschritt auf der Stelle.

Sie laufen auf der Stelle.

Höher das Knie, höher! Habt ihr lausigen Mitwisser euch endlich besonnen, wer euer Schuldiger ist? Wer meldet sich?

Schweigen.

Wer meldet den anderen?

Schweigen.

MÜNSTERER Gnade Gott, wer es getan hat.

FELDWEBEL Gruppe fünf macht heute Nachtdienst. Keine Arbeitspause. Wir arbeiten in einem Zug durch. Und immer noch Laufschritt an Ort und Stelle, marsch, marsch.

9. Bild

Von der Decke senkt sich ein Bett herab, das in der Luft schweben bleibt, sonst bleibt die Bühne leer.[56]*Hinter der Bühne wird mit ruppigen Soldatenstimmen gesprochen:*
Eidesstattliche Erklärung Kreszenz Pichler
Eidesstattliche Erklärung Hansi Mittermeier
Eidesstattliche Erklärung Karoline Perger
Eidesstattliche Erklärung Luise Bachl
Eidesstattliche Erklärung Maria Motzer
Eidesstattliche Erklärung Berta Haberer
Eidesstattliche Erklärung Paula Vogelsang

STIMME DES FELDWEBELS Schert euch zum Teufel! [57]

10. Bild
Luitpoldpark

Ferne Parademusik. Sonntäglich aufgeputzte Bürger und Pioniere begegnen sich auf verschlungenen Wegen, stehen beisammen und trennen sich wieder, es bilden sich neue Begegnungen.

I

Feldwebel sieht Alma auf sich zukommen und möchte am liebsten in einen Seitenweg einbiegen, aber sie stellt ihn vorher.

FELDWEBEL Der schönste Tag ist mir verdorben, weil meine Pioniere alle ein Alibi haben. Da ist jeder ein Aal, ich fasse keinen am Griff. Und jetzt kommt mir die Katze über den Weg.

ALMA Ist das nicht der Hochstapler, der nur ein Trinkgeld gezahlt hat? Sie erinnern sich wohl nicht so gern?

FELDWEBEL Fräulein, Sie müssen sich irren.

ALMA Wen ich einmal sehe, den weiß ich.

FELDWEBEL Fräulein, ich tue Ihnen ja nichts ab. Sie verwechseln mich aber.

ALMA Hinterher wollen Sie es nicht gewesen sein.

FELDWEBEL Pech. Soldaten sehen alle gleich aus unter der Mütze. Da ist nicht viel Unterschied drin.

ALMA Für mich schon. Sie schulden mir genau soviel, wie Sie bezahlten. Angeblich hatte der Herr nicht mehr dabei.

FELDWEBEL Sie sind zu hartnäckig, Fräulein.

ALMA Haben Sie es heute dabei?

FELDWEBEL Nein. Wenn einer nicht herausrücken will, was können Sie machen?

ALMA Das haben wir abgesprochen zuvor, Sie. Da bin ich im Recht.

FELDWEBEL Pech. Ich streite es ab. Mit mir war nichts.

ALMA Ich verlange nur, was wir ausgemacht haben.

FELDWEBEL Sie haben auch was davon gehabt. Sie können sich nicht beklagen.

ALMA Für mich ist es die Frauenfrage.[58] Entweder ich bin es Ihnen wert –

FELDWEBEL Fräulein, hinterher nie!

ALMA Schuft.

FELDWEBEL Da müssen Sie schon mit Fotos arbeiten, wenn Sie einen erpressen.

ALMA Pfui.

FELDWEBEL Sie müssen sich schon einen Kumpel beibiegen, der das für Sie besorgt.

ALMA Ich will Sie nicht verstehen, mein Herr.

FELDWEBEL Nämlich, Sie wissen nicht, was Sie noch für eine Anfängerin sind.

ALMA Sie kommen sich wohl sehr überlegen vor?

FELDWEBEL Es macht sich. Ich habe die Erfahrung.

ALMA Sie sind ja auch älter.

FELDWEBEL Das stimmt. Ich bin nicht mehr so grün.

ALMA *Hohn:* Armer Mann! Was müssen Sie für ein armer Mann sein.

FELDWEBEL Feine Dame.[59] F D. *Feldwebel ab.*

ALMA Was essen müssen ist ja noch keine Schande. *Alma ab.*

2

Berta und Münsterer

BERTA Er hat gesagt, er wart bei der Bank, und jetzt kommt er nicht.

Münsterer pirscht sich heran.

MÜNSTERER Heißen Sie Berta?

BERTA Sie haben bei mir nichts zu suchen.

MÜNSTERER Bei der Bank kann ein jeder stehn.

BERTA Aber ich habe da was ausgemacht und Sie nicht.

MÜNSTERER Wer sagt Ihnen denn, daß ich nicht auch was ausgemacht habe?

BERTA Ich war zuerst da.

MÜNSTERER *auf sie zu, sie weicht ihm aus:* Wir brauchen uns da nichts vormachen. Das habe ich bald gespannt, wenn eine gern möchte. Da bin ich dann nicht so.

BERTA Unterstehen Sie sich!

MÜNSTERER Fräulein, kommen Sie her, ich tue Ihnen nichts. Mein militärisches Ehrenwort. Da bin ich groß.

BERTA Auf einmal –

MÜNSTERER Einen schönen Gruß vom Korl, und er kann heut nicht kommen.

BERTA Warum sagen Sie das jetzt erst?

MÜNSTERER Ich muß immer herausbringen, wie ich auf ein Fräulein wirke mit der Wirkung. Das ist bei mir instinktiv.

BERTA Schön ist, wenn einer so an wen denkt.

MÜNSTERER Lieber wär er mir nicht so schön. Ich an seiner Stelle wär hergangen.

BERTA Sie brauchen ihn mir nicht schlecht machen.

MÜNSTERER Gestern war er mit Ihnen zusammen, heut treibt er es mit einer anderen. Drankommen muß eine jede.

BERTA Von wem reden Sie überhaupt?

MÜNSTERER Und da wollen also Sie seine Kinder aufziehn?

BERTA Der hat kein Kind.

MÜNSTERER Der hat sogar eine Braut.

BERTA Das ist nicht der gleiche Korl.

MÜNSTERER Korl Lettner, gibt nur den einen. Der hat die Kinder nur so herumsitzen in den Städten.

BERTA Das glaube ich nicht.

MÜNSTERER Ich werde doch nicht einen Kameraden hinhängen bei seinem Mädel.

BERTA Sie sind kein Kamerad. Sie wollen es ihm bloß geben.

MÜNSTERER Sie halten sich wohl für die einzige Flamme?

Berta schweigt.

Ich kann ja nicht riechen, daß er Ihnen verheimlicht, was an ihm alles dranhängt. Ich hab gemeint, da hat er Ihnen schon lang was gesagt.

BERTA Der sagt mir alles.

MÜNSTERER Warum hat er Sie dann versetzt?

BERTA Er wird seinen Grund haben.

MÜNSTERER Daran müssen Sie sich schon gewöhnen, Sie sind nur eine von vielen. Und Sie müssen noch zulernen, Fräulein.
Ab.

BERTA Da ist man nicht geachtet.[60] *Ab.*

3

Pionier, Alma

PIONIER Fräulein, haben Sie Zeit für einen armen Soldaten?

ALMA Ich bin nicht mehr frei. Sie verstehn.

PIONIER Schade. Wir treffen uns noch.

ALMA Sie haben mich nicht verstanden. Ich bin nicht mehr frei. Bei mir kann sich ja auch was verändern.

PIONIER Sie hören auf? Ich halte das für einen Verlust. Die anderen sind ja Gänse.
Pionier ab.

4

Einige Mädchen haben es aus der Ferne beobachtet. Zwei kommen heran.

ERSTE Die Alma markiert da so eine.

ZWEITE Nämlich die Mädchen lassen dir sagen, daß es nicht mehr schön ist, wie du dich anschmeißt.

ALMA Sie sollen aufpassen, daß sie sich nicht selber anschmeißen. Anschmeißen habe ich nicht nötig. Ich habe eben meine Verehrer.

ZWEITE Dann haben wir auch Verehrer, wenn wir es machen wie du.

ERSTE Bilde dir bloß nicht ein, daß du so schön bist.

ALMA Es reicht für meine Zwecke, Fräulein.

ZWEITE Deine Zwecke kennen wir. Das sind mir so Zwecke.

ALMA Ihr seid wohl die ganz Heiligen?

ZWEITE Das ist der Unterschied. Wir wissen, wie weit wir gehn.

ALMA So? Bist du nicht gestern abend mit dem mit den Blattern am

Kaiserwall gegangen?

Dem zweiten Mädchen verschlägt es die Rede.

ERSTE Vielleicht warst du selber am Kaiserwall. Du hättest sie sonst nicht gesehn.

ALMA *zur Zweiten:* Mit dem Kerl bin ich in einer halben Stunde weiter wie du, das kann ich dir sagen.

ZWEITE Das möchte ich vorerst bezweifeln.

ERSTE Du bist mir schon die richtige Marke.[61] An der Hecke hat sie Geld genommen.

ALMA Wann habe ich Geld genommen?

ERSTE Bitte, das haben wir gesehn.

ALMA Mein Freund hat es mir aufgedrängt. Die Soldaten wollen mich glatt so haben.

ERSTE Mir hat noch keiner was aufgedrängt.

ALMA Du hast eben nicht die Nachfrage. Du bist ja auch nicht intelligent.

ERSTE Bitte sehr, ich bin intelligent.

ALMA Im übrigen täuscht ihr euch alle.

MÄDCHEN Ph!

Die Mädchen weichen zurück. Alma ab nach der anderen Seite.

5

Berta, Korl.

BERTA Dann kann ich dir gar nicht helfen?

KORL Nicht, wenn du dich anhängst.

BERTA Ich will dir nicht aufsitzen.

KORL Heut muß ein Mädel sich was gefallen lassen, weil es zuwenig Mannsbilder gibt.

BERTA Es gibt mich und gibt dich.

KORL Dir fehlt bloß die Übersicht.[62] Nämlich der Frauenüberschuß, der ist zu groß.

BERTA Es gibt mich und gibt dich, und andere brauchen wir nicht.

KORL Du sollst mich nicht immer so lieben, das macht mich noch rasend.

BERTA Wenn ich nicht anders kann.

KORL Weil du nicht schlau bist.

BERTA Wie wird man schlau?

KORL Wenn man von einem Mann was will, darf man nicht zeigen, was er mit einem machen kann.

BERTA Wie soll ich denn sein?

KORL Hast du schon einmal was gehört, daß man sich nicht an einem Menschen einhalten kann? Es hebt nicht.

BERTA Es würde schon heben. Du willst nicht tragen.

KORL Den ganzen Tag muß ich mich schikanieren lassen, bei den Weibern lasse ich mich aus. Das muß eine einsehn.

BERTA Ja dann –

KORL Braucht eine bloß mitspielen.

BERTA Wenn du was zum Waschen hast, bring es mir her. Ich steh auf in der Nacht. Ich geh ins Waschhaus.

KORL Der Mensch muß nicht alles haben. Das machen wir schon im Dienst.

BERTA Was kann ich für dich tun?

KORL Mich einen freien Mann sein lassen.

BERTA Weißt du, was daraus entsteht?

KORL Das macht doch mir nichts. Da hättest du dich eben früher rar machen müssen.

BERTA Wenn ein Mann ein Mädel so weit hat und sie schaut keinen anderen an, dann fragt er einmal, ob sie ihn heiraten will, weil dann geheiratet wird.

KORL So schaust du aus. Schon wäre einer drin in der Falle.[63]

BERTA Du kannst doch jetzt nicht so sein.

KORL Mit mir nicht. Ich kann das nicht brauchen.

BERTA Du hast mich verraten.

KORL An uns muß man glauben. Dann muß man sich von uns verraten lassen. Dann darf man weinen, wenn man mag, und dann muß man erst recht an uns glauben.

BERTA So lang, bis eine kaputt ist.

KORL Die Richtigen halten es aus.

BERTA Aber das mußt du doch spüren, wer du für mich bist.

KORL Einen Fetzen muß man aus euch machen.[64] *Ab.*

6

Fabian, Bibrich.

FABIAN Da steht eine und kann nicht heimgehn.

BIBRICH Was hat man mit dem Mädel angestellt? Das ist doch sonst ein ganz braves Mädel. Die muß ja verrückt werden. Da tun sie mit ihr herum. Dann ist nie wer da.

BERTA Man kann sich nicht einhalten. Es hebt nicht.[65]

FABIAN Fräulein Berta, Sie sehn doch, daß sich der andere nichts daraus macht.

BERTA Aber das ist doch ganz anders. Der weiß es bloß nicht.

BIBRICH So einer will es nicht wissen.

FABIAN Sie werden den Menschen nicht noch verteidigen.

BERTA Aber das habe ich doch tief in mir drin. Und ich mache nichts falsch. Mich wird einer noch brauchen, das weiß ich.

BIBRICH Hören Sie auf, sich was vorzumachen.

FABIAN Der ist nicht zu helfen.

BIBRICH Laß sie laufen. Das ist eine, die ist ganz umstellt.

Beide ab, sie läuft weg.

7

Alma, Korl.

ALMA Wie habe ich das gemacht?

KORL Was gemacht?

ALMA Vor dem Bierzelt. Mensch. Sie werden mich noch kennen. Ich habe Sie diesem nachgemachten Napoleon aus den Klauen gerissen.[66]

KORL Sie waren der lockere Vogel?

ALMA Und was sind Sie für ein Vogel?

KORL Auch nicht besser. Man lebt.

ALMA Man wünscht sich zuviel.

KORL Kann schon sein.

ALMA Der Napoleon hat Sie wohl dick?

KORL Wir haben ihn dick. Die ganze Mannschaft weiß, daß es nicht mehr weitergeht mit dem komischen Josef. Nur – beim

Barras geht das immer weiter.

ALMA Man muß doch was machen können.

KORL Beim Barras passiert schon mal was, aber den Richtigen trifft es nie. Das müßte schon dumm gehn. [67]

ALMA Und keiner kommt, der ihn umlegt.

KORL Das geht nicht. Das ist eingebaut, daß es nicht geht. Das ist von Anfang an drin.

ALMA Sie müssen es wissen.

KORL Beim Barras sind sie hinter lauter Vorschriften verschanzt, man kann nie heran. Man kann sich nur beschweren, das kann man. Lieber nicht. Da hält einer besser den Mund.

ALMA Wenn sich alle zusammentun?

KORL Alle zusammen, das gibt es nicht. Man kann gar nichts machen. Der bleibt uns.

ALMA Das ist mir schon der richtige Josef. Spielt den Lebemann, bleibt hinterher alles schuldig. Er hat mich hineingelegt, Sie verstehen.

KORL Von ihm war das ein Fehler. Ich würde den Fehler nicht begehn. Ich verspreche nie was. Ich halte noch weniger, als ich verspreche.

ALMA Ja, aber Sie sagen es gleich. Das ist ein Unterschied.

KORL Sie machen es wohl noch nicht lang?[68]

ALMA Und ich hör auch schon wieder auf . Da ist kein Leben für mich. Ich bin menschlich enttäuscht. Die Herren befriedigen bloß ihre Sinnlichkeit.

KORL Haben Sie das nicht gewußt?

ALMA Zuvor war ich bei einer Madam. Ich geh nie wieder zu einer Madam.

KORL Das kann ich mir denken.

ALMA Ich hab mich ins Freie gewagt, aber dort war es nicht frei.

KORL Wo es viele sind, das ist doch nur Ersatz. Was das Interessante ist, das lassen die weg.

ALMA Ich lerne noch.

KORL Das hör ich gern.

ALMA Ich bin eben unvorsichtig, das ist das Ganze.

KORL Für mich ist das nicht so schlimm. Die ersten Tritte sind nicht gleich die richtigen Tritte. Das macht nichts.

ALMA Und ich bin hineingesprungen und springe auch wieder heraus.

KORL Das nenne ich Mut.

ALMA Ich bin nur noch nicht im richtigen Zug.

KORL Der richtige Zug kommt nicht immer.

ALMA Der kommt, das weiß ich. Mir träumt immer, daß es mich hebt. Eins und zwei und spring! Ich könnte den ganzen Tag springen. Sie auch?

KORL Das war einmal.

ALMA Mir hilft alles voran, daran glaube ich. Und ich geh nach Berlin. Und ich will das Leben, wo es mich herumschmeißt und packt.[69]

KORL Am liebsten ginge ich mit.

ALMA Sie sehen es mir vielleicht nicht an, aber in mir ist was, das schiebt mich durch dick und dünn. Ich brauche nur noch den Dreh, der mich durchzieht.

KORL Sie werden sich schon was anlachen.

ALMA Da ist mir nicht angst. Aber es muß so sein, daß sich was rührt. Ich wäre sonst überhaupt kein Mensch.

KORL Wie wäre es mit Arbeit?

ALMA Für mich kann Arbeit ein Dreh sein, das glauben Sie nicht?

KORL Ich kann mir was Schöneres denken.

ALMA Zum Beispiel?

KORL Sie gefallen mir, Mädchen. *Er umfaßt sie.*

11. Bild
Straße Nähe Donau

Es wird dunkel. Roßkopf und Münsterer in etwas angetrunkenem Zustand. Sie rollen eine Tonne vor sich her.

MÜNSTERER Was ein schlauer Soldat ist, der verrollt sich in seine Falle und ist gut aufgehoben, wenn er nicht mehr weiß, was er tut.

ROSSKOPF Aber wir wissen es noch.

MÜNSTERER Beinahe. Zwick mich, und ich werde nüchtern.

ROSSKOPF Ich zwicke nie.

MÜNSTERER Idiot!

ROSSKOPF Und ich will nicht aufgehoben sein, ich will meine Freiheit.

MÜNSTERER Für uns gibts keine Freiheit. Außer bei unserem Mädel, und das ist halb so wild.

ROSSKOPF Jeder lausige Zivilist hat es besser als ich. Darum wische ich heut einem Zivilisten eins aus.

MÜNSTERER Ja, du wischst ihm eins aus.

ROSSKOPF Wenn ich selber dran glauben muß, warum soll der Zivilist nicht dran glauben? Logisch. *Er betrachtet überrascht die Tonne.* Wo ist das Ding wieder her?

MÜNSTERER Das hast du geklaut.

ROSSKOPF *beleidigt:* Re-quiriert.

MÜNSTERER Du hast sie mitgehen lassen.

ROSSKOPF Du behandelst mich wie einen Dieb. Horneule, du bist zu streng.

MÜNSTERER Ich bin nicht streng.

ROSSKOPF Horneule, roll nicht immer dahin. Da geht es in die Kaserne.

MÜNSTERER Du weißt schon lang nicht mehr, wo es in die Kaserne geht.

ROSSKOPF Die Kaserne hängt mir zum Hals heraus. Ich verlange meine per – sönliche Freiheit.

MÜNSTERER Du verlangst deine persönliche Freiheit.

ROSSKOPF Ich will meinen Spaß haben, verdammt, jetzt werde ich wild. Horneule, machst du mit bei einem Spaß?

MÜNSTERER Dann will ich aber auch lachen.

ROSSKOPF Du wirst lachen. Du wirst vielleicht nicht gleich lachen, später im Bett wirst du lachen.

MÜNSTERER Ich weiß nicht. Wie willst du es machen?

ROSSKOPF Ich werde einen Zivilisten fressen. Ich werde ihn nicht echt fressen, meine Tonne soll ihn fressen. Und dann mache ich mit ihm, was ich will.

MÜNSTERER Nämlich?

ROSSKOPF Wir fangen uns den dahinten. Ich werde es dir zeigen.

MÜNSTERER Unblutig, Max.

ROSSKOPF Unblutig, Horneule.

MÜNSTERER Dann lasse ich dir deinen Lauf.[70]

Fabian nähert sich.

ROSSKOPF Da geht er und weiß von nichts.

MÜNSTERER Er ist noch nicht trocken hinter den Ohren.[71]

ROSSKOPF Egal.

Sie passen ihn ab und umstellen ihn.

Was glaubst du, was wir da haben?

FABIAN Eine Tonne.

ROSSKOPF Das ist keine gewöhnliche Tonne. Was glaubst du, wer in der Tonne drin ist? – Ein nacktes Weib.

FABIAN Das hättet ihr wohl gern.

ROSSKOPF Es ist eine gewisse Berta, wenn du sie kennst.

FABIAN Das könnte euch so passen.

ROSSKOPF Du brauchst es nicht glauben.

Die Tonne steht senkrecht, Fabian guckt hinein, im selben Augenblick nehmen sie ihn an den Füßen und kippen ihn kopfüber in die Tonne. Sie schlagen den Deckel zu.

Bist du schön weich gefallen auf dein nacktes Weib? Liegst du schon auf ihr drauf? Du konntest es wohl nicht lassen?

FABIAN Aufmachen! Sofort aufmachen!

ROSSKOPF Hilft nichts. Jetzt bist du schon drin.

FABIAN Ich schreie. Hilfe! Überfall!

MÜNSTERER Das wirst du nicht tun. Dir geht es schlecht, wenn du schreist.

ROSSKOPF Schön brav sein, Zivilist.

Fabian ist verstummt.

MÜNSTERER Wer schon drin ist, der ist verratzt. Das mußt du dir merken.

ROSSKOPF Du mußt stillhalten, Zivilist. Du bist stumm, oder wir machen dich stumm. Jetzt hast du die Wahl.

FABIAN Bitte, laßt mich heraus.

MÜNSTERER Es ist nur Spaß, Zivilist.

ROSSKOPF Ja, wenn wir Ernst machen würden. Aber wir machen nicht Ernst.

MÜNSTERER Wir spielen nur. Spielst du mit?

ROSSKOPF Wir werden Krieg spielen.

MÜNSTERER Und das mußt du freiwillig tun. In den Krieg schlittern wir immer freiwillig hinein.

ROSSKOPF Gegen den Krieg läßt sich nichts machen.

MÜNSTERER Wir werden dir zeigen, wer du im Krieg bist. Zivilist, paß auf.

Sie kippen die Tonne um und stoßen sie.

Wenn der Krieg kommt, dann bist du eine Laus.

ROSSKOPF Für mich bist du jetzt schon eine Laus. *Stößt die Tonne zu Münsterer.*

MÜNSTERER Ich bin auch eine Laus, aber das wirst du nie sehn. *Stößt die Tonne zu Roßkopf.*

ROSSKOPF Was machst du in einem Krieg? Du hast Schiß und sonst nichts.

MÜNSTERER Nur keine Aufregung. Das haben wir alle.

ROSSKOPF Hast du Angst?

Fabian schweigt.

Nein? Hast du immer noch keine Angst?

MÜNSTERER Er will es bloß nicht zugeben. Er lügt.

FABIAN Hört auf. Ich habe was falsch gemacht. Laßt mich heraus.

MÜNSTERER Einen, der was falsch gemacht hat, können wir nicht herauslassen.

FABIAN Was verlangt ihr von mir, daß ihr aufmacht?

ROSSKOPF Ich werd verrückt, der Mann zahlt.

MÜNSTERER Aber das geht nicht mit Geld. Das wäre einfach.

ROSSKOPF Warum nicht?

MÜNSTERER Wir wissen es anders. Sag uns, wer unser Holz

gestohlen hat, dann lassen wir dich aus.

ROSSKOPF Horneule, du bist ein Genie. Antwort da drinnen!

FABIAN Ich war es nicht.

MÜNSTERER Wir waren es auch nicht. Aber uns hängt man es an, und uns triezt man dafür.

ROSSKOPF Wer war es dann?

FABIAN Ich weiß es nicht.

MÜNSTERER Weißt du es wirklich nicht? Dann machen wir weiter. *Er stößt.*

FABIAN Aufhören.

MÜNSTERER Du willst es nur nicht gewesen sein. Sing, Vogel, sing.

ROSSKOPF Wir zum Beispiel haben es uns nicht unter den Nagel gerissen, verdammt.

FABIAN Ich weiß nicht, ich weiß nicht, ich weiß nicht. Ihr dürft so was nicht machen.

MÜNSTERER *Kniebeuge:* Vergib uns, Kleiner. Wir wissen nicht, was wir tun.[72]

FABIAN Woher soll ich was wissen, ich war nicht dabei. Genausogut warst es du selber.

MÜNSTERER Da wären wir wieder.

ROSSKOPF Du bringst es nicht aus ihm heraus. Er war nicht dabei.

MÜNSTERER Der Kerl hat unverschämtes Glück, daß wir ihm glauben. *Er macht den Deckel auf.*

FABIAN Warum tut ihr mir weh? Was habe ich euch denn getan?

ROSSKOPF Du bist ein feiner Max, du wirst nicht geschlaucht. Wenn du deine faule Zeit hast, werden wir schon geschlaucht.

FABIAN Von mir doch nicht.

ROSSKOPF Das ist gleich. Ich bin nachtragend.

Jedesmal wenn Fabian herausklettern will, drückt er ihn wieder hinunter.

Du bist auch so einer, den es nichts angeht. Aber es geht dich was an. Zum Teufel! An der Eskaladierwand geschunden, mit voller Ausrüstung und Gepäck durchs Gelände gehetzt.

MÜNSTERER – ganze Kilometer durch nasse Äcker und Straßen gekrochen.

ROSSKOPF – und das alles geht dich nichts an. Aber das wär grad für dich was, und das lasse ich jetzt an dir aus. *Er schlägt den*

Deckel wieder zu.

FABIAN Das ist feig. Ihr seid feig.
Sie rollen ihn auf eine größere Distanz.
Mörder! Ihr seid Mörder. Mörder!

ROSSKOPF Wenn du das sagst, rolle ich dich an die Donau und werfe dich da hinein.

MÜNSTERER Unblutig, Max! Du hast genug gehabt. Laß ihn gehn. *Er zieht Roßkopf weg.* Es regnet. *Zu Fabian in der Tonne.* Du bist im Trockenen. Jetzt bist du fein heraus. *Sie gehen weg, Fabian öffnet den Deckel.*

12. Bild
Donau

Starker Wind. Fabian allein am Ufer. Rudergeräusche eines Pionierbootes. Man hört die Stimmen vom Boot.

FELDWEBEL Eins zwei. Eins zwei. Eins zwei. Ruder stoppt. Anker abrollen lassen.

JÄGER *wiederholt:* Anker abrollen lassen.

Man hört das Abrollen der Ankerwinde und dann einen Schrei. Fabian erschrickt.

JÄGER Mann über Bord.

Das Seil hat noch weiter abgerollt und steht dann still.

KORL Sitzenbleiben! Das ist zu gefährlich.

MÜNSTERER Der Feld hängt mit dem Fuß in der Schlinge. Von allein kommt er nicht frei.

KORL Er ist so gut wie tot.

JÄGER Aber man muß doch was tun. Man muß ihn aus der Schlinge schneiden. Da muß einer hinunter.

KORL Hast du soviel Lunge, dann machs. Ich hab nicht soviel Lunge.

MÜNSTERER Das kann keiner. Das Seil ist zu zäh. Der Wellengang ist zu stark.

KORL Sitzenbleiben, verdammt. Das Boot legt sich zu stark auf die Seite.

MÜNSTERER Das Boot ist am Kippen. Das Seil zieht uns hinunter.

KORL Kappt das Seil. So kappt doch endlich. Kappt das Seil.

MÜNSTERER Wir müssen noch alle versaufen.

Das Seil wird gekappt.

FABIAN Die Schufte lassen ihn drunten.

KORL Zum Ufer und nichts wie raus. Eins zwei. Eins zwei.

Sie rudern ans Ufer. Ein paar springen heraus und ziehen das Boot an Land. Alle springen heraus. Sie sind abgekämpft.

FABIAN Was macht einer für ein Gesicht, der so was getan hat?

KORL Wir haben gar nichts getan.

FABIAN Das war zuwenig. Sie sind beobachtet worden.

KORL Na und?

FABIAN So nah am Ufer einen absaufen lassen.

KORL Dafür braucht man keine Entfernung.

JÄGER Wir hätten ja getaucht, aber wir konnten nicht tauchen. Der Wellengang war zu stark.

FABIAN Ihr hättet es gleich machen müssen, solang das Seil noch gespannt war.

KORL Sie waren schön im Trockenen, was?

FABIAN Ihr hättet den Mann nicht einmal suchen müssen. Das Seil hätte euch hingeführt zu dem Mann. Statt dessen kappt ihr das Seil.

KORL Wir hätten, wir hätten! Dann wären wir auch noch versoffen. Das Seil zog uns ja schon hinunter.

FABIAN Man wird euch nicht glauben, daß ihr den Mann nicht umgebracht habt. Der Tote hat einen Zeugen.

KORL Eine Untersuchung gibt das immer. Aber mit keinem Zivil.

FABIAN Der Mann ist mit dem Fuß nach hinten in eine Schlinge getreten, da hat ihn das Seil kopfüber in die Donau gerissen und er war der Gefangene vom Seil.

KORL Das muß er sich selber zuschreiben. Das war Zufall. Wir haben dem Mann keine Schlinge gelegt.

FABIAN Ihr habt nur an euch gedacht, wie es passiert war. Die eigene Haut war euch lieber.

KORL Das ist nicht strafbar. Wir waren selber zu nah daran. Es hätte mehr Menschenleben gekostet. Was wollen Sie überhaupt? Es war ein Zufall. Den nimmt man uns ab.

MÜNSTERER Es ist eben passiert. Passieren kann immer was. Wir können uns das Wasser nicht aussuchen.

KORL Wenn wir über einen Fluß müssen im Krieg, suchen wir uns das Wasser nicht aus.

JÄGER Wenn wir bei jedem Unfall den Kopf verlieren, könnten wir den harten Dienst gar nicht machen.

KORL Es war ein Unfall. Das ist sogar die Wahrheit, und man wird uns verhören, und das sagen wir aus.

FABIAN Ihr redet mir zuviel.

JÄGER Es war ein Unfall, den nimmt man uns ab.

KORL Zeigen Sie uns doch an, wenn Sie ein Maul dafür haben. Wir haben es auch.

FABIAN Aber ihr habt nicht das Letzte versucht.

KORL Das Letzte versuchen, entschuldigen Sie schon, für wen? Wir hätten uns ja für den Mann umbringen können. Wir halten das für übertrieben.

MÜNSTERER Es geht eben hart auf hart.[73]

JÄGER Der Mann ist eben ertrunken.

KORL Bedanken Sie sich bei der Donau.

FABIAN Ein Maul habt ihr wie ein Schwert. Na, was gehts mich eigentlich an? Ich bin nicht die Wehrmacht.

MÜNSTERER Der Mann hats überstanden.

KORL Der Mann bleibt schon nicht drunten.

MÜNSTERER Warten Sie es ab, bis wir einen anderen Wellengang haben.

KORL So einen Anker läßt man nicht hinten. Mit dem Anker, wenn der gehoben wird, kommt der Mann von allein wieder herauf .

13. Bild
Luitpoldpark

Alma, Fabian.

ALMA Ist das wahr, daß die Soldaten Sie überfallen haben?

FABIAN Die haben mich nicht bloß überfallen. Die haben mich am Leben bedroht.

ALMA Das sind doch nette Menschen. Ich kann das von denen nicht glauben.

FABIAN Aber manchmal wird es Ernst.

ALMA Aber Sie leben noch.

FABIAN Das werfen Sie mir noch vor.

ALMA Ich will Ihnen nichts wegtun.

FABIAN Sie tun mir nichts weg. Mir tut keine mehr was weg. Ich war ja verbohrt, daß ich mich an eine hingehängt habe, die mir gar nichts will. Damit ist es aus.

ALMA Das haben Sie doch nicht nötig.

FABIAN Ich renne nicht mehr um mein Leben, wenn ich mir aus einer was mache. Das geht auch so.

ALMA Es geht sogar besser.

FABIAN Warum sich erst lange in ein Unrecht setzen, nur damit der andere recht hat? Warum freiwillig eine Vorgabe machen? Ich bin schon auch noch was wert. Wie du mir, so ich dir.

ALMA Ich halte das nur für gesund.

FABIAN So ist es doch. Einer will dem anderen Herr werden.[74] Wer wird den anderen fressen?

ALMA Das ist oft gar nicht so schön. – War es das erste Mädel, an das Sie sich hingehängt haben?

FABIAN Eine muß die erste sein. – Aber da war nichts. Ich kam an die nicht heran.

ALMA Dann war es nicht viel.

FABIAN Und das wird ihr noch leid tun. Ich könnte mir vorstellen – *schweigt.*

ALMA Was?

FABIAN An Ihnen hätte ich mehr.

ALMA Ich habe mir immer einen gewünscht, bei dem ich die erste war. Ich komme immer nur an die anderen hin, ich weiß nicht warum.

FABIAN Jetzt werden Sie nicht komisch.

ALMA Eine muß die erste sein.

FABIAN Klar.

Sie gehen ins Gebüsch.

14. Bild
Brückenbaustelle

Die Brücke ist fast fertig. Pioniere machen letzte Handgriffe. Es wird mit Scheinwerfern gearbeitet.

ROSSKOPF Morgen wird die Brücke eingeweiht. Aber da sind wir schon nicht mehr hier.

MÜNSTERER Immer sind wir im Vortrupp.

JÄGER Parole: Ab nach Küstrin.

MÜNSTERER Ich vergesse schon, daß wir hier waren. Ich bin schon gar nicht mehr da.

ROSSKOPF Eine Stadt ist wie die andere. Die Feldwebel sind überall gleich.

MÜNSTERER Das ist wie Ameisen im Hirn. In der Luft liegt Aufbruch.

ROSSKOPF Inzwischen sind unsere Mädchen schon wieder neu.

JÄGER Es kommt immer was nach.

MÜNSTERER Aber es sind immer die gleichen Flüsse. Am frühen Morgen ist die Luft immer rauh.

ROSSKOPF Ein Gemeiner bleibt ein Gemeiner, und ein Offizier bleibt Trumpf, und er sticht.

MÜNSTERER Wir werden keine Herren beim Barras. Wir werden bloß gewitzter.

BERTA *tritt auf:* Ich muß einen sprechen, den Korl Lettner.

ROSSKOPF Korl, ein Mädel will dir was.

KORL Das werden wir gleich haben. Wo ist der Feld?
Sie deuten weit nach hinten.
Macht keine Geschichten, ich trete aus. *Er geht zu Berta hinüber.* Jetzt läufst du mir doch wieder nach.

BERTA Ja. Aber das ist nicht so leicht. Man merkt, wo man nicht auskann.

KORL Muß die Hochzeit gleich sein?

BERTA Wem seine Hochzeit?[75] Ich weiß schon, ich schinde keine heraus.

KORL Die Geschichten kenne ich. Da bist du nicht die erste und wirst auch nicht die letzte sein. Da bist du bei mir falsch.

BERTA Ich glaube nicht mehr daran. – Ich kann doch nicht von dir lassen.

KORL Das mußt du wissen.

BERTA Jetzt hast du mich mit Haut und Haar.[76]

KORL Von mir aus können wir gleich was haben. Ich bin dann nicht so.

BERTA Doch nicht hier. Ich habe mich frei gemacht für die Nacht.

KORL Tut mir leid, ich bin beim Militär. Ich kann nur auf einen Sprung weg. Wenn der Feld kommt, muß er mich eingereiht sehn.

BERTA Mir wird schlecht.

KORL Das will ich alles nicht wissen.

Er schlägt sich mit ihr in ein Gebüsch. Die Pioniere arbeiten weiter, es dauert eine gewisse Zeit. Sie singen: Was nützet mir ein schönes Mä – ädchen, wenn andere drin spazierengehn?[77]

ROSSKOPF *schadenfroh:* Scheinwerfer nach links.

Berta und Korl kommen durch den Busch ins Licht. Die Pioniere johlen und pfeifen.

KORL Nehmt euer kindisches Licht weg, verdammt.

ROSSKOPF Was willst du, es ist bloß der Neid.

MÜNSTERER Ach wo, das machen wir doch jeden Tag.

Sie nehmen das Licht weg.

KORL Denk nicht mehr daran. Diese Männer sind morgen schon fort. Steh auf jetzt. Nimm dich zusammen. Andere müssen es auch.

Berta steht auf.

BERTA War das alles?

KORL Warum? Hat dir was gefehlt?

BERTA Wir haben was ausgelassen, was wichtig ist. Die Liebe haben wir ausgelassen.

KORL Eine Liebe muß keine dabei sein.

BERTA Das ist mir jetzt ganz arg.

KORL Berta, ich muß mich einreihn. Du kannst hier nicht bleiben. Du gehst jetzt am besten weg.

BERTA Ich kann es nicht. So kann es nicht aus sein. Warum sind die Männer morgen schon fort?

KORL Berta, ich habe es dir bis jetzt nicht gesagt, für uns ist

Abmarsch. Wir gehn diese Nacht noch zurück nach Küstrin.
BERTA Man muß mir doch Zeit lassen.
KORL Wir sind im Vortrupp. Wir sind immer die ersten.
BERTA Das geht doch nicht. Ich bin damit noch nicht fertig.
KORL Das mußt du abschneiden, Berta. Einfach abschneiden. Andere müssen es auch.
BERTA Aber ich kann so nicht leben.
KORL Du wirst müssen.
Bunny macht den Vogelpfiff. Korl läuft zu den anderen und reiht sich ein.
Fotograf nähert sich mit klappendem Werbegeräusch.
FOTOGRAF Meine Herren, Sie ziehen jetzt fort aus dieser geschichtlichen Stadt. Sie haben uns diese feste Brücke gebaut, damit uns an Sie ein dauerndes Andenken bleibt. Sicher werden auch Sie den Wunsch verspüren, daß Sie ein kleines Souvenir an den Aufenthalt der Pioniere in Ingolstadt mitnehmen. Das Bild beschwert Sie nicht, es findet in jeder Brieftasche Platz und es kostet fast gar nichts, in der Gruppe pro Mann nur drei Mark. Sie können das Bild hervorholen in jedem freien Augenblick –
MÜNSTERER Denkste –
FOTOGRAF Sie können es Ihren diversen Bräuten und Ihren treusorgenden Eltern zeigen.
ROSSKOPF Das nehmen wir auch noch mit, Mann. Das lassen wir uns nicht entgehn.
FOTOGRAF So, stellen Sie sich auf, meine Herren, Sie gruppieren sich zwanglos, die großen Herren nach hinten –
Pioniere stellen sich auf.
Die hinteren stehen, die mittleren knien, die vorderen liegen, damit ich alle ins Bild bringe, damit jeder Charakterkopf draufkommt. Schauen Sie nicht in den Apparat.
Er knipst.
Ich bitte um Angabe der werten Adressen, ich kassiere sofort, ich entwickle noch heut.
MÜNSTERER Hier ist die Sammeladresse der Einheit. Sie brauchen nur noch die Namen, wer zahlt. Zur Kasse, wer sein Gruppenbild haben will.
PIONIER Jäger – *zahlt*

FOTOGRAF Jäger – *schreibt*
PIONIER Pfaller – *zahlt*
FOTOGRAF Pfaller – *schreibt*
PIONIER Angerer – *zahlt*
FOTOGRAF Angerer – *schreibt*
PIONIER Gensberger – *zahlt*
FOTOGRAF Gensberger – *schreibt*
PIONIER Bachschneider – *zahlt*
FOTOGRAF Bachschneider –*schreibt*
PIONIER Roßkopf – *zahlt*
FOTOGRAF Roßkopf – *schreibt*
KORL Lettner – *zahlt*
FOTOGRAF Lettner – *schreibt*
MÜNSTERER Münsterer – *zahlt*

Er ist der letzte.

FOTOGRAF Münsterer – *schreibt*

Ich bedanke mich, Herr Münsterer. Ich habe die Namen, Sie haben die Quittung für alle. *Er gibt ihm den Durchschlag der Sammelquittung.* Ich stehe auch für Einzelfotos gern zur Verfügung. Sechs Bilder in Kabinettform kosten nur zwanzig Mark. Vom Unteroffizier abwärts die Hälfte, und wer ist schon Unteroffizier.

ROSSKOPF Na, Lettner, du wirst dich verewigen lassen mit deiner Verflossenen, soviel Anstand wirst du noch haben. Dann hat sie dich auf dem Papier.

KORL Komm, Berta. *Er stellt sich mit ihr in Positur.* Sie müssen das Bild an die Dame schicken.

BERTA Berta Kobold, hier, Dollstraße siebzehn.

FOTOGRAF Ich kassiere sofort.

Berta will Geld herausholen.

KORL Sie kassieren bei mir. *Er zahlt.*

MÜNSTERER Achtung.

Der nächste Feldwebel tritt auf. Er ist auch nur eine neue Auflage des ertrunkenen Feldwebels.

Melde: Brücke fertig zur Übergabe.

FELDWEBEL Ich nehme ab. *Er inspiziert die Brücke.* Antreten in Marschordnung.

Die Pioniere stellen sich zum Abmarsch auf.
Wir alle wissen, es geht zurück nach Küstrin. Ich erwarte von meiner Mannschaft ein vorbildliches Verhalten und daß mir keine Klagen aus der Bevölkerung kommen. Der Soldat muß wissen, daß er als Staatsbürger in Uniform immer im Blickfeld der Öffentlichkeit steht. Vor dem Verlassen der Kaserne prüft der Soldat seine Uniform. Die Taschen sind zugeknöpft, die Schuhe blank, die Mütze sitzt gerade und ist ohne Kniffe.
Es ist ungehörig, in Trupps den Gehsteig zu blockieren und andere Personen an den Straßenrand zu drängen. Singen und auffälliges Benehmen unterbleibt. Rauchen auf der Straße ist unsoldatisch. Betrunkenen, Aufläufen und Schlägereien geht der Soldat aus dem Weg. Bei der Auswahl von Lokalen ist er vorsichtig, vor den Eingängen steht er nicht herum. Ausschweifende Tänze passen nicht zur Uniform. Politische Versammlungen darf der Soldat in Uniform nicht besuchen. Das sind Heeresvorschriften, das muß euch durch Mark und Bein gehn, da gibts keine Lockerung, da gibts kein eigenmächtiges Verhalten. Jedes Zuwiderhandeln wird disziplinarisch bestraft.
Fertig zum Abmarsch. Links schwenkt, im Gleichschritt – Marsch!
Die Pioniere marschieren über die Bühne.
Wir singen: zicke zacke, zicke zacke, hoi, hoi, hoi![78]
Die Pioniere singen.

Notes to the text

Pioniere: engineers, sappers. Military units responsible for the construction of roads and bridges, and the clearing of mines, to facilitate troop movements.

1 **oh du schöner Westerwald:** traditional marching song, still popular today ('Heute wollen wir marschier'n,einen neuen Marsch probier'n, in dem schönen Westerwald, ja da pfeift der Wind so kalt').

2 **Du hast doch den Sohn von deiner Herrschaft:** it is regarded as quite natural that Berta, as a 'Dienstmädchen', might be involved in a sexual relationship with her employer or his eldest son(see also 39 and 65). In a letter to Rainer Roth of 28 December 1972 Fleißer wrote of the status of the housemaid: 'Von der Außenwelt waren die Dienstmädchen ziemlich abgesperrt, weshalb sie neugierig waren. Das Zusammenleben mit heranwachsenden Söhnen in einem Haus schuf oft die Gelegenheit, die sie anderswo nicht finden konnten und für die Söhne war es bequem, daß sie an ihnen ihre ersten Erfahrungen machten. Das wurde offiziell zwar verschwiegen, aber doch vielfach praktiziert. Die Ehemänner hatten da schon mehr die Ehefrau zu scheuen, bei einem Witwer war das schon anders.' Fleißer treats this theme in her story *Stunde der Magd*. See also Stritzke, 34 f.

3 **kennen:** used here in the sense of 'wissen', common in South German speech.

4 **eventunell:** for 'eventuell', 'maybe'. In her attempt to impress Berta with sophisticated language, Alma makes a mistake.

5 **Da geht den Großen wieder was hinaus:** (S. German) 'Typical, the big shots have managed it (i.e. got something for nothing) again.'

6 **In deinem Alter ... lang:** 'When I was your age I managed that kind of thing (i.e. found a girl) easily.'

7 **Augen auf ... schon noch:** 'Keep your eyes open and your chin up, and don't get emotionally involved. You'll manage it.'

8 **Der kann man was mucken:** 'You can have your own way with her.'

9 **Hauptsächlich ... charakteristisch:** Zeck uses bombastic and rather clichéd formulations in his advice to Fabian.

10 **Küchenmädchenlied:** Well-known example of this form of folk-song, popular with 'Dienstmädchen' and attested in various versions. Elisabeth Bond-Pablé, co-translator of the play for the 1991 London production, suspects the work of Brecht in the modified (and ungrammatical) second line, which often reads 'aus Berlin'. In an interview (*Mat.*, 347), Fleißer points out that she was the first to use this song in a play, suggesting that Brecht borrowed it from her for his *Schweyk im Zweiten Weltkrieg* (*Gesammelte Werke*, 1, 775). It appears, though, that Brecht knew the song before he met Fleißer (see Kässens/Töteberg, '"fast schon ein Auftrag von Brecht" ...', 110).

11 **kennen:** here for 'kennenlernen'.

12 **Das sieht ... haben:** 'You don't look as if you can afford to be choosy.'

13 **Der Ärgste ... nachgibt:** 'It's a fool who doesn't give in', possibly adapted from 'der Klügere gibt nach'.

14 **bist:** the verb 'sitzen' is commonly used with the auxiliary 'sein' in S. German.
15 **Die Mädel ... haben:** 'Girls always think they can take the mickey out of/deceive you.'
16 **Das gwöhnst:** S. German colloquial form of 'du gewöhnst dich daran', 'you'll get used to it'.
17 **wie eine Beschwörung:** in her desperate attempt to fend off Korl's persistent sexual advances, Berta rhetorically invokes a higher authority. Korl rejects her attempt to cleanse ('verklären', 'to cleanse, transfigure') him.
18 **Ich schrei ... aus:** 'I'm shouting my head off.'
19 **Sie hat ... auf:** 'She's got her head screwed on.'
20 **Einmal ... hin:** 'Once bitten, twice shy.'
21 **Heißt ... zieht:** 'If it's for life, they're not interested.'
22 **Frechwerden ... zu:** 'You're not old enough yet to start getting cheeky.'
23 **Die Englein ... singen:** play on 'die Engel singen hören', 'to be in great pain.'
24 **Auf ... nicht:** 'The town authorities are deaf in that ear.'
25 **am Hut stecken:** for 'an den Hut stecken', 'to stuff it.'
26 **Der Kerl ... blind:** 'The fellow is blind in that eye', i.e. 'he doesn't want to see' (that a superior is present).
27 **Maß:** the noun 'Maß', meaning 'measure', is neuter in standard German. The S. German feminine form means a one-litre tankard of beer.
28 **die taufrische Begleitung:** ironic/mock chivalrous term for female companion.
29 **der Zorn ... Sie:** 'it's you she's angry with'.
30 **Das würde ... lassen:** 'No one would dare say that to me, even if I was asleep.'
31 **Stolz weht die Flagge Schwarz–Weiß–Rot:** patriotic navy song by Richard Thiele (1883), especially popular around the time of the First World War ('Stolz weht die Flagge Schwarz–Weiß–Rot von unsres Schiffes Mast; dem Feinde Tod, der sie bedroht, der diese Farben haßt').
32 **Auf der Welt:** usually 'auf die Welt', a toast. The substitution of the dative for the accusative is not uncommon in S. German, e.g. 'am Hut stecken', 49.
33 **Die Letzten ... sein:** Matthew 19: 30; Mark 10: 31; Luke 13: 30. Uttered by Fabian here as a conventional response to the sergeant's toast, though not without relevance in the context of their discussion of oppression.
34 **mein ... sagen:** 'a little bird is trying to tell me something'.
35 **Ich habe ... Holz:** 'I heard some rumour about stolen wood.'
36 **Den Kopf ... herunter:** 'I'll rip their heads off.'
37 **Es sticht ... muß:** 'I have an inkling of who must have done it.'
38 **Die Stadt ... blasen:** 'The town will tell you where to get off.'
39 **Holzhackergesellen:** woodcutter's song, a folk-song still popular in various versions in Bavaria today e.g. 'Mir sein de lustigen Holzhackergeselln und mir taen, was ma wölln' (origin unknown). Fleißer refers to it in her autobiographical sketch *Frühe Begegnung* as the kind of simple, 'gestic' song which so fascinated Brecht (II, 299).
40 **Ich muß ... Mädchen:** 'I need a piss, girl.'
41 **Ich will ... hat:** 'I want to see a real salute.' The sergeant dismisses Korl's greeting as 'not even a pathetic attempt at a regulation salute', demands a more respectful and energetic acknowledgement ('hineinwerfen', 'spritzen'), and threatens to force him to salute properly ('Feuer unter dem Arsch').
42 **Die Fliegen ... kriegen:** According to information provided by Klaus Gültig, this song might well be a version of the 'Vogelhochzeit' kind of ditty ('Schnader-

hüpferl'), possibly a variation on Karl Valentin's animal alphabet (F: 'die Fliege, die Fliege, saß draußen auf der Stiege ...').

43 **Ich ... verdächtig:** The sergeant, assuming that Korl is paying for Berta's attentions, wonders where the money might have come from. He is shocked to learn that (unlike himself) Korl is sufficiently successful with the opposite sex not to require the services of a prostitute.

44 ***hat daran schwer zu schlucken*:** 'is at a loss for words'.

45 **Nicht ... früher:** The sergeant replies to Alma's flirtatious enquiry with a sexual innuendo which sets the tone for their subsequent exchange and ensuing relationship. Alma soon introduces the question of payment for her attentions ('es gibt eine Kasse').

46 **In deine ... auch:** Berta adopts the role – and diction – of the romantic lover, using popular clichés to express her attraction to Korl. He is left cold by such sentimentality.

47 **Wir sind ... Militär:** 'We're completely unimportant compared to the soldiers.'

48 **Davon ... gefressen:** 'It wears you down completely.'

49 **geschützt:** 'protected' (by law). As a 'Dienstmädchen' Berta has certain rights. Unertl warns her not to complain to the authorities about his treatment of her.

50 **schon ... Tempel:** '... and she's gone/disappeared' (possibly derived from John 2: 15).

51 **Sie hat ... Schwert:** 'She has a sharp tongue.'

52 **Dann weht ... Wind:** 'Then things will be very different.'

53 **jetzt ... dran:** 'Now we're in for it.'

54 **Wir haben ... Gegend:** 'We spend enough time out here (i.e. at the building site) during the day.'

55 **Sie steht ... hier:** 'We've had it up to here.'

56 ***Von der Decke ...*:** This rather unnaturalistic stage effect is not typical of the play as a whole. It demonstrates, of course, where encounters between the soldiers and the local girls take place.

57 **Schert ... Teufel:** 'Go to hell.'

58 **die Frauenfrage:** 'the women's question' i.e. the issue of women's rights. Typically, Alma uses a concept which is hardly appropriate to the context.

59 **Feine Dame:** 'prostitute'.

60 **Da ist ... geachtet:** 'I'm not respected here.'

61 **Du bist ... Marke:** 'You're a fine one'.

62 **Dir fehlt ... übersicht:** 'You don't understand the situation.'

63 **So schaust ... Falle:** 'You'd like that, wouldn't you? Then I'd be trapped.'

64 **Einen Fetzen muß man aus euch machen:** 'You need ripping to shreds.' This is reputed to have been a favourite expression of Brecht's in the 1920s (see I, 447). According to Fleißer, Korl has some of the features of the early Brecht (I, 447 and letter to Rainer Roth, 21 June 1971).

65 **Man kann ... nicht:** Berta quotes Korl's words from the previous scene. She is clearly much preoccupied with his rejection of her serious intentions.

66 **Ich habe ... gerissen:** 'I saved you from that twopenny-halfpenny dictator (the sergeant).'

67 **Das müßte ... gehen:** 'There's no chance of that happening.'

68 **Sie machen ... lang?:** 'You haven't been working as a prostitute for long, have you?'

69 **Und ich ... packt:** 'I want some real action from life.' Alma is attracted by the

excitement of the metropolis as an escape from the monotony and claustrophobia of the provinces.

70 **Dann ... Lauf:** 'Then I'll give you free rein.'

71 **Er ist ... Ohren:** 'He's still wet behind the ears.'

72 ***Kniebeuge:* ... tun:** ironic reference to Luke 23: 34: 'forgive them; for they know not what they do.'

73 **Es geht ... hart:** 'It's a hard life.'

74 **Einer will ... werden:** 'Everyone wants to get the upper hand.'

75 **Wem ... Hochzeit?:** 'Whose wedding?' Korl suggests that Berta is trying to pressurise him into marriage. By this stage she seems to have given up any such aspirations, although she continues to insist on the necessity of emotional commitment.

76 **Jetzt ... Haar:** 'Now I'm yours body and soul.' Another of Berta's romantic clichés.

77 **Was nützet mir ... spazierengehn?:** Obscene song, origin unknown, revealing in this context the dual moral standards of the military. On the one hand they expect their girls to be generous with physical favours, on the other they acknowledge (here humorously) that promiscuity undermines a girl's value.

78 **zicke zacke, zicke zacke, hoi, hoi, hoi!:** common chant among military and other male groups.

A note on the three versions of the text

Three versions of *Pioniere* exist. The following is a brief summary of the principal differences between them:

1 The original version, written 1926/27, produced at the Komödie Dresden on 25 March 1928, unpublished (typescript in the Marieluise-Fleißer-Archiv, Ingolstadt). Contains the essence of the Berta–Korl (here Karl) plot, and something of the tensions apparent among the military and in the Unertl (here Benke) household. Fabian is the bitter rival of the soldiers for the attention of the local girls. He tries to injure Karl by tampering with a ladder – it is the sergeant who is the actual victim – and attempts, unsuccessfully, to bribe the sergeant into providing dynamite for an attack on the bridge. Fabian does, however, manage to outwit a visitor to Ingolstadt in the matter of the purchase of a car and, having failed to secure the attentions of Berta, enters into a liaison with Alma. The play ends with the ceremonial opening of the bridge.

2a The text of the scandalous Berlin production (Theater am Schiffbauerdamm, 30 March 1929), much influenced and amended by Brecht, now extant only as a fragment (Marieluise-Fleißer-Archiv). As a result of criticisms of the Dresden performance, Fleißer tightened up her text considerably. The car sub-plot and other distracting elements were omitted, and Fleißer's original songs were replaced. There is a clearer focus on the Berta–Karl relationship (subtitle: 'Soldaten und Dienstmädchen'). At Brecht's insistence, Fleißer injected 'Pfeffer' into her text: she added a scene in which three grammar-school boys discuss the female anatomy, set the exchange between Korl and the 'Dienstmädchen' Frieda on syphilis (present in the first version) between gravestones in a cemetery, and permitted Brecht to stage Korl's deflowering of Berta in a large crate which rocked rhythmically to and fro on stage. Unlike the original version, the play ends with a speech by the sergeant.

2b The censored Berlin version, cleared by the police, first performed on 31 March 1929 and on forty-two subsequent occasions (published in I, 187-222). Provides the basis for Faßbinder's stage and television adaptations. Largely identical with 2a, to which rapid (overnight) changes were made in order to gain police consent for further performances. The 'spice' added at Brecht's insistence, specifically those elements outlined above, was removed.

3 The 1968 revision (the 'authorised version' printed here), first performed at the Residenztheater Munich on 1 March 1970, published in *Theater heute* (8/1968) and, with minor modifications, in *Spectaculum* 13 (1970), in I, 127-85, and in the paperback *Ingolstädter Stücke* (71-129). A thorough revision and expansion of the text, which is now twice as long as 2. Two major episodes were added, both intended to reinforce tensions and aggressions: the swimmers with their theft of the wood, and the death of the sergeant. Overall the social–critical dimension of the text is consciously intensified, primarily by the development of the Unertl scenes, and the theme of oppression and frustration (in both civilian and military spheres) is brought out more clearly. Fabian has a quite different role, now appearing more passive. New names are used, and the replacement sergeant introduced. A new speech is provided at the end, based on military regulations.

Arbeitsteil

1: Fragen zum Textverständnis

1. Bild:

(1) Was erfahren wir in diesem Gespräch über das Verhältnis zwischen Berta und Alma? Was will Berta von Alma? Warum ist Berta so unsicher? Worin zeigt sich das Unabhängigkeitsstreben Almas? Welche Bedenken hat Berta hinsichtlich eines Verhältnisses mit dem 'Sohn ihrer Herrschaft'?

(2) Wie definiert Fabian seine momentane Situation? Was sagt er über Berta? Welchen Rat gibt Zeck? Was für ein Frauenbild hat Zeck? Was für eine Wirkung haben Zecks Worte auf Fabian?

(3) Was für eine Funktion hat das 'Küchenmädchenlied' im Text? Warum läßt Alma Berta stehen? Wie verläuft das Gespräch zwischen Alma, Roßkopf und Münsterer? Wie verhält sich Alma gegenüber Jäger? Warum?

(4) Fassen Sie die wichtigsten Phasen dieses Gesprächs zusammen. Wie verhält sich Korl am Anfang dieser Szene? Welche Positionen vertreten die beiden Geprächspartner? Warum hält Berta ihren eigenen Namen für so wichtig? Warum wird Korl handgreiflich? Was erwartet er von Berta? Wie reagiert er auf seine Enttäuschung? Wie reagiert Berta am Ende der Szene? Wer gewinnt zum Schluß die Oberhand, und wie?

2. Bild:

Was für einen Eindruck gewinnen wir von Unertl am Anfang der Szene? Was verlangt er von einem Dienstmädchen? Wie verteidigt Fabian Berta? Warum behauptet Unertl, nicht heiraten zu wollen? Was erwartet er von einer möglichen Ehepartnerin? Wie hat er seinen früheren Heiratsantrag an seine Verkäuferin formuliert? Was schließt er aus ihrer Ablehnung? Wie sieht er seine Zukunft? Was für eine Meinung hat Fabian von seinem Vater?

3. Bild:

Wie zieht Zeck Fabian auf? Was passiert Bibrich? Wie reagieren die anderen? Warum fühlen sich die Schwimmer benachteiligt? Wo müssen sie schwimmen? Welchen Plan schlägt Zeck vor?

4. Bild:
Wie ist diese Szene aufgebaut? Wie schätzen Sie den Stand der Beziehung zwischen Berta und Fabian zu diesem Zeitpunkt ein? Warum muß Berta zu Korl hinübergehen? Was sagt Korl über die Verhältnisse beim Militär? Welchen 'Verdruß' hat der Feldwebel? Was kann er dagegen tun?
Wie beabsichtigt er, in dieser Sache vorzugehen? Welche Konsequenzen wird die Situation für ihn haben? Wie und warum schikaniert der Feldwebel Korl? Wie benimmt sich Alma gegenüber dem Feldwebel? Welche Bedingung stellt Alma für Intimverkehr mit dem Feldwebel? Wie weit hat sich die Beziehung zwischen Berta und Korl inzwischen entwickelt? Welche Warnung spricht Korl aus? Warum wird er so aggressiv? Wie gehen sie am Ende der Szene auseinander?

5. Bild:
Welche Funktion hat diese Szene? Wie verhalten sich die Schwimmer gegenüber dem Polizisten?

6. Bild:
Was hat Korl hier vor? Aus welchem Grund? Welche Konsequenzen wird Korls Tat laut Bunny haben?

7. Bild:
Was wirft Unertl Berta vor? Wie verteidigt sie sich? Auf welcher Seite steht Fabian? Worüber beschwert sich Berta? Welche Unterstellungen macht Unertl? Womit droht er? Gelingt es ihm, Berta zu unterdrücken? Wie ist das Verhältnis zwischen Vater und Sohn am Ende dieser Szene?

8. Bild:
Fassen Sie die Handlung dieser Szene zusammen. Was fordert der Feldwebel am Anfang der Szene? Warum verdächtigt er sofort Korl und die Gruppe fünf? Wie verhalten sich die Pioniere einander gegenüber nach dem Anschlag? Wie läßt der Feldwebel seine Frustration an den Pionieren aus?

9. Bild:
Welche Funktion hat diese kurze Szene? Was ist der Sinn der Regieanweisung?

10. Bild:
(1) Worüber streiten sich Alma und der Feldwebel? Was meint Alma mit der 'Frauenfrage'? Warum ist für sie der Feldwebel 'ein armer Mann'? Wie rechtfertigt Alma ihr Verhalten?

(2) Was soll Münsterer Berta ausrichten? Warum sagt er nicht gleich, was er auszurichten hat? Was erzählt er über Korl? Was muß Berta 'zulernen'?

(3) Was ist die Aussage dieser kurzen Episode?

(4) Was werfen die Mädchen Alma vor? Wie verteidigt sie sich? Wodurch unterscheidet sich Alma von den anderen Mädchen? Was für eine Einstellung hat Alma ihnen gegenüber?

(5) Wie verläuft dieses Gespräch? Welche 'übersicht' fehlt Berta? Warum ist sie nicht so schlau? Wie definiert Korl das Verhältnis zwischen den Geschlechtern? Was will er von Berta?

(6) An welcher überzeugung hält Berta trotz allem fest?

(7) Wie definiert Korl die Strukturen und Zustände beim Militär? Warum ist Alma 'menschlich enttäuscht'?
Welche Ideale und Hoffnungen hat sie noch? Was erwartet sie von Berlin? Wie endet diese Szene?

11. Bild:
Was verübeln Roßkopf und Münsterer dem Zivilisten? Wie locken sie Fabian in ihre Falle? Warum bezeichnet Roßkopf Münsterer als 'Genie'? Warum ist Fabian 'ein feiner Max'? Was läßt Roßkopf an Fabian aus? Wie beschreibt Roßkopf das Leben beim Militär?

12. Bild:
Was passiert dem Feldwebel? Wie? Warum lassen die Pioniere es zu? Was wirft Fabian ihnen vor? Wie rechtfertigen die Pioniere ihr Verhalten?

13. Bild:
'Damit ist es aus': was meint Fabian damit? Wie wird er sich in Zukunft verhalten? Was findet er an Alma attraktiv? Was reizt sie an ihm? Welche Lektion haben die beiden im Laufe des Dramas gelernt?

14. Bild:
Was sagen die Pioniere am Anfang der Szene über ihren Aufenthalt in Ingolstadt? Welches Zugeständnis macht Berta jetzt an Korl? Wie und unter welchen Umständen findet der Geschlechtsakt statt? Warum ist Berta danach so enttäuscht?
Womit ist Berta 'noch nicht fertig'? Warum schenkt Korl Berta ein Bild? Was sind Sinn und Funktion der Schlußrede des neuen Feldwebels?

Diskussionsthemen
(s. hierzu 'Introduction' und Arbeitsteil 2: Materialien)

'Die Komödie spielt 1926' (36): worin zeigt sich das historisch Spezifische des Dramas?

Inwiefern ist Ihrer Meinung nach die Thematik des Werks heute noch relevant?

'Die nackte Notwehr' (39): inwiefern sind Zecks Worte relevant für die Darstellung zwischenmenschlicher Beziehungen überhaupt in diesem Drama?

Inwiefern bietet das Stück eine Analyse der Provinzmentalität?

Welche Manifestationen und Auffassungen von Liebe sind in diesem Text evident?

'Ingolstadt steht für viele Städte' (*Mat.*, 353): Inwiefern trifft das zu?

Untersuchen Sie die Bedeutung der Sexualität in Marieluise Fleißers Text.

Analysieren Sie das Verhältnis zwischen Fabian, Berta und Korl.

Besprechen Sie die Rolle und das Bild der Frau in dem Stück.

Welche Funktion haben die (in der Endfassung neu hinzugekommenen) Szenen im Haushalt Unertl (2., 7. Bild)?

Untersuchen Sie die Funktion der 'Tonnenszene'(11. Bild).

Inwiefern läßt sich *Pioniere* als 'Volksstück' bezeichnen?

Untersuchen Sie die Form des Dramas und ihre Angemessenheit in bezug auf die Thematik.

'Zentrales Thema des ... Stücks ist das an ihrer Sprachnot und emotionalen Unfähigkeit scheiternde Bedürfnis "kleiner Leute", sich anderen mitzuteilen und so über die trostlosen Verhältnisse "hinausgehoben" zu werden' (Klaus Hübner). Stimmen Sie dieser Interpretation des Dramas zu?

'Verspottung der Provinz; Ingolstadt als idiotisches Nest; Soldaten als

Schweinehunde' (Erik Krünes). Ist diese Sicht des Dramas zulässig?

Inwiefern demonstriert dieses Drama "die Bedingtheit des privaten Verhaltens durch die soziale Stellung" (Peter Schaarschmidt)?

'Aus heutiger Sicht ist weder die Empörung der Ingolstädter Bürger über die vermeintliche "Nestbeschmutzerin" noch die Begeisterung Brechts über diese trostlosen, zuweilen allzu locker gefügten Genrebilder aus dem Bürger- und Soldatenleben so recht verständlich' (Michael Schmidt). Nehmen Sie Stellung hierzu.

'Heute lesen wir *Pioniere* als ein Stück, in dem die Pervertierung der Liebe durch hierarchische Gesellschaftsstrukturen bloßgelegt wird' (W. Kässens/M. Töteberg). Ist dieser Auffassung zuzustimmen?

'Die Vielseitigkeit [des] Motivs des gesellschaftsimmanenten Drucks macht das Stück heute noch relevant .... Alle Hauptfiguren, Alma, Berta, Fabian, Korl, zeigen Sehnsüchte nach einem anderen Leben, doch sämtliche Befreiungsversuche aus den Systemzwängen scheitern' (Moray McGowan). Nehmen Sie Stellung dazu.

'die Verbindung von Wirklichkeit und Poesie, das Zurücktreten von Handlung zugunsten der sinnlichen und sozialen Dimension, die Tendenz zu einer ausstellenden Art von Darbietung' (Günther Rühle). Trifft diese Einschätzung von Marieluise Fleißers Technik Ihrer Meinung nach zu?

'Die distanzierte Sicht, die kühle Beobachtung von Wirklichkeit kennzeichnet das Stück. Zustände werden geschildert, ohne daß dem Zuschauer eine Interpretation mitgeliefert wird' (W. Kässens/M. Töteberg). Untersuchen Sie die Gültigkeit dieser These.

2: Materialien

Kontext

'Ingolstadt, das war der Angelpunkt ihres Lebens und das Generalthema des Werkes der Marieluise Fleißer. Die alte baierische Residenz- und Universitätsstadt an der Donau, die sich seit der Gegenreformation als Bollwerk des Katholizismus versteht, war für die Fleißer ein Synonym für kleinstädtische Beengtheit und Provinzmief, für Bigotterie und Heuchelei, vor allem aber für die Unterdrückung von Außenseitern. Dazu kam noch, daß die Stadt im 19. Jahrhundert und bis zum Ersten Weltkrieg vom Militär geprägt war. Denn nachdem bereits im 15. Jahrhundert die Residenz und zu Beginn des 19.

Jahrhunderts auch noch die Universität nach Landshut verlegt worden waren, wurde zwischen 1827 und 1848 Ingolstadt zur Festung und zur Garnisonsstadt, der zweitgrößten in Bayern, ausgebaut, um die Stadt vor dem Absinken in die Bedeutungslosigkeit zu retten. Das Militär wurde wesentlicher Wirtschaftsfaktor und bestimmte das Leben in dieser Stadt. Als dann nach dem Ersten Weltkrieg aufgrund der Bestimmungen des Versailler Vertrages Militär und Wirtschaftssegen ausblieben, entstand aufgrund dieser Beeinträchtigung der finanziellen Einkünfte und damit auch des bürgerlichen Selbstbewußtseins – stärker als anderswo – ein Klima aus Nationalismus, Revanchismus und unkritischer Glorifizierung der "goldenen Zeiten" vor dem Ersten Weltkrieg. Muß es da noch verwundern, wenn in dieser Atmosphäre Außenseiter, die – wie die Fleißer – die enge Welt solch einer Kleinstadt aufzeigen, als "Nestbeschmutzer" gebrandmarkt werden?' (Macher, 544)

(*a*) Welche Rolle spielte das Militär in der Geschichte Ingolstadts?
(*b*) Wozu führte der Abzug des Militärs aus der Stadt?
(*c*) Wie definiert Macher die Atmosphäre in Ingolstadt zur Zeit Marieluise Fleißers?
(*d*) Inwiefern ist diese Atmosphäre Ihrer Meinung nach in *Pioniere* evident?

'**Militarismus**, ein in Frankreich um 1860 aufgekommenes Schlagwort, mit dem von liberaler Seite militärische Eigenarten und Forderungen in Friedenszeiten auf polemische Weise charakterisiert wurden; im weiteren Sinn bezeichnet der Ausdruck das Vorherrschen militärischer Formen, Denkweisen und Zielsetzungen in Staat, Politik und Gesellschaft. Während Verteidigungsbereitschaft, Wehrpflicht, soldatisches Ethos, militärische Disziplin in allen zivilisierten Staaten als Notwendigkeit anerkannt werden, haben sich in der Geschichte zu allen Zeiten Staatstypen herausgebildet, deren Organisation durch die Vorherrschaft militärisch-kriegerischer Prinzipien geprägt wurde. In der neuesten Geschichte wird dafür der Ausdruck Militarismus gebraucht.' (*Der große Brockhaus*, 1971)

(*a*) Wie wird der Begriff 'Militarismus' hier definiert?
(*b*) Welche Anzeichen von Militarismus lassen sich in *Pioniere* erkennen?
(*c*) Läßt sich das Drama Ihrer Meinung nach als 'anti-militaristisch' bezeichnen? (s. hierzu die Texte zur Berliner Aufführung 1929).

Selbstzeugnis Marieluise Fleißers

'Ich wünsche, daß mich vor allem die jungen Menschen hören, sehn und lesen und daß sie durch mich einen Einblick bekommen in das, was hinter der Oberfläche steckt. Darüber hinaus schreibe ich für alle Aufgeschlossenen, die bereit sind, den Druck und die Ungerechtigkeit zu erkennen im Alltäglichen, im gar nicht einmal so Seltenen, in Wirklichkeit Perfiden. Ich will ihnen den Blick dafür öffnen, was anders sein müßte. Ich lege Verletzungen bloß, die geheilt werden müßten. Ich habe keine wirkliche Hoffnung, sie zu heilen. Es wäre schön, könnte es nützlich sein, wenn man erkennt. Ich lasse mich nur zu gern überraschen. Die Erfahrung ist anders .... Ich schreibe für jene, die entschlossen sind, zu erkennen. Ich schreibe für jene, die sich nichts vormachen lassen.' (Marieluise Fleißer,'Schreiben – für wen?', IV, 522)

(*a*) Was beabsichtigt die Fleißer mit ihrem Schreiben?
(*b*) Glaubt sie, mit ihren Werken die Welt verändern zu können?
(*c*) Trifft diese Aussage Ihrer Meinung nach auch für Inhalt und Technik von *Pioniere* zu?

Rezeption und Wirkung
Zur Berliner Aufführung 1929

'Ein Volksstück ist geplant, ein Soldatenstück – von einer Frau, die erstens keinen Funken von Instinkt für wirkliches Volkswesen hat und überdies als Frau bei einem Thema wie diesem fehl am Platze ist. Die Welt des Militärs, der richtigen Männer, ist eine sehr ordentliche, anständige Welt – auch wenn es dort sehr wenig literarisch und dafür etwas derb und animalisch zugeht. Es ist eine Welt für Männer, gestaltbar nur durch Männer – einer Frau unzugänglich, eben weil es die Welt ohne Weiblichkeit ist. Die Verfasserin, die den üblichen ohnmächtig literarischen Haß der Schwachen auf dieses Reich der Kraft hat, versucht trotzdem, diesem maskulinen Bezirk beizukommen. Das Ergebnis ist Katastrophe, literarisch wie menschlich.' (Paul Fechter, *Deutsche Allgemeine Zeitung*, 3 April 1929)

(*a*) Wie wird die Welt des Militärs in dieser Kritik dargestellt?
(*b*) Warum ist die Fleißer laut Fechter bei diesem Thema 'fehl am Platze'? Welches Bild der Frau impliziert dieses Argument?
(*c*) Was verstehen Sie unter dem Begriff 'wirkliches Volkswesen'?

'Das alte Militär! Welch Bühnenwitz für abgedankte Demokraten! Wer sieht nicht das neue Militär, das Militär der Hundertschaften, das Militär mit Sowjetstern und Hakenkreuz? Dieses Militär wird euch eure literarischen Nacktspässe schon vertreiben und nicht mit sich spaßen lassen. Dieses Militär wird Politik machen, gleich welche. Anstelle der schmierigen Sexualmanöver werden andere Manöver treten.

Die Dinge sind am Explodieren. Gegen diese Zensur der Gemeinheit, unter der die Nation leidet, wird sich eine andere rechtschaffene Zensur durchsetzen, die aus ihrer Gesinnung keinen Hehl machen wird und auch gemein werden kann, wenn es nicht mehr anders geht und die Schändung des Berliner Theaters zum Himmel schreit. Dann wird sich eine Literatin wie Marieluise Fleißer nicht mehr vor einem Pfeifkonzert lächelnd verbeugen können, dann werden andere Mittel Zensur sein.' (Dr Richard Biedrzynski, *Deütsche Zeitung*, 2 April 1929)

(*a*) Was bedeutet 'das Militär der Hundertschaften, das Militär mit Sowjetstern und Hakenkreuz'?
(*b*) Welche Drohung wird hier ausgesprochen?
(*c*) Erläutern Sie die Begriffe 'rechtschaffene Zensur' und 'Gesinnung'. Welche politische Einstellung wird hier evident?

'Ein trauriger Fall höchst sittlicher Korruption. Ein marxistisch-barmatiadischer Fall. Jüdische Geilheit und Dirnentum beschmutzen bairisches und deutsches Gemüt .... Das Theater am Schiff-bauerdamm hat bekannt politischen Einschlag und darum ist es auch möglich, daß dieses perverse und zotische Stück mit seinem antimilitärischen Charakter dort als Erstaufführung erscheinen konnte .... Eine Scheidung der Geister tut not. Vor allem aber anläßlich dieses Falles die Erkenntnis, daß es mit unverstandener Toleranz unmöglich ist, der sittlichen Bedrohung Einhalt zu gebieten .... Hier ist eiserne Energie nötig, um diesen Augiasstall auszurotten. Der Kampf der Nationalsozialisten gegen die jüdisch-marxistische Volkspest hat sich durch diesen Skandal neuerdings als dringende Notwendigkeit erwiesen.' (Be., *Donaubote*, 5 April 1929)

(*a*) Worin manifestiert sich die nationalsozialistische Einstellung in dieser Passage?
(*b*) Warum wird Marieluise Fleißers Stück so vehement angegriffen, und inwieweit ist diese Kritik berechtigt?

'Stets sprechen da einfache, wirre, gehemmte Kleinstadtmenschen: Kleinbürger und Proletarier. Sie sprechen einfach, wirr, gehemmt, aber in so natürlicher, so ihre tiefste Menschlichkeit enthüllender Art ... daß eben diese Art, Menschen zu offenbaren, zugleich die Ursprünglichkeit und Stärke dieser Dichterin offenbart. Der Marieluise Fleißer gab ein Gott zu sagen, nicht was sie leidet, sondern was armselige, weltferne, junge Menschen leiden, die nicht selber ihr Leid und ihre Lust ausdrücken können ... es gibt da Gespräche zwischen Dienstmädchen und Soldaten, die in ihrer tastenden Sehnsucht, in ihrer schlagenden, urwüchsigen Kraft einzigartig sind in unserer Dramatik .... Und deshalb darf dieses Stück nicht verboten werden ....' (Kurt Pinthus, *8-Uhr-Abendblatt*, Berlin, 2. April 1929)

(*a*) Worin besteht laut Pinthus die Stärke der Fleißer?
(*b*) Identifizieren Sie im Dramentext Dialogstellen, die Pinthus' Argument belegen bzw. widerlegen.

Zur Aufführung der Neufassung 1970

'Die sozialen Voraussetzungen der 1929 im Berliner Theater am Schiffbauerdamm ausgepfiffenen Komödie *Pioniere in Ingolstadt* haben sich so radikal geändert, daß es scheint, als wäre das Stück von der Droste oder von der Sappho .... Denn von der Ausbeutung irgendwelcher Dienstmädchen oder abhängiger Hilfskräfte kann heute – ein Blick in den Anzeigenteil jeder Tageszeitung genügt – so direkt gewiß nicht mehr die Rede sein. Und was die Strenge des Militärs betrifft, so stehen ... auch anscheinend die bürgerlichen Staaten aus welchen Gründen auch immer hinter den totalitären zurück: der "gemeine Soldat" kann, wenn nicht alles täuscht, in der Bundesrepublik massivere Rechte geltend machen als in der DDR, in Rußland und wahrscheinlich auch in China. Unsere zeitgenössischen "Abhängigkeiten" und "Versklavungen" sehen anders aus.' (Joachim Kaiser, *Süddeutsche Zeitung*, 3. March 1970)

(*a*) Fassen Sie Kaisers Vorbehalte gegenüber dem Stück zusammen.

'Es scheint, als ob in diesen [Kaisers] Sätzen die reaktionäre Interpretation der "sozialen Voraussetzungen" und ihrer Veränderung unmittelbar zusammenhängen mit einem grundlegenden Mißverständnis dessen, was an dem auf dem Theater Gezeigten "aktuell" zu sein habe. Ästhetisch reaktionär ist daran die Zumutung, auf dem Theater müßten zumindest die sozialen Verhältnisse mit jenen, die der Rezensent um sich herum erlebt oder durchschaut, komparabel sein. Wäre es an dem, müßte sich niemand um einen anspruchsvolleren Begriff von geschichtlicher Erfahrung, als er zur Zeit dem Theater abverlangt wird, bemühen. Was aber ist tatsächlich komparabel zwischen Theater und tagtäglichem Leben? Kaum die demonstrierten Sachverhalte, wohl eher deren Ideologien, welche diese nachweislich überdauern.

Das Stück der Fleißer erzählt knapp skizzierte Passagen aus der Entfremdungsgeschichte des bürgerlichen Individuums, führt die einfachen emotionalen und mentalen Verkümmerungen, Verstörungen von kleinen Leuten vor, deren privates Glücksbedürfnis inmitten eines in repressiven Beziehungen geregelten gesellschaftlichen Umgangs nur mehr als verschwommene, kitschige Innerlichkeit sich auszudrücken weiß ....' (Botho Strauß, *Theater heute*, April 1970).

(*a*) Wie versucht Strauß, Kaisers Argumente zu widerlegen?
(*b*) Was findet er an Kaisers Kritik reaktionär, und warum?

(*c*) Was meint er mit einem 'anspruchsvolleren Begriff von geschichtlicher Erfahrung', und worin sieht er die heutige Relevanz des Dramas? Suchen Sie Belege für diese Interpretation im Text.

Franz Xaver Kroetz zu Pioniere in Ingolstadt

'Brutalität wird sichtbar gemacht durch den Ausstellungscharakter der Fleißerschen Sprache. Mit Brecht hat diese Sprache nichts zu tun. Haben die Proletarier Brechts immer einen Sprachfundus zur Verfügung, der ihnen de facto nicht zugestanden wird von den Herrschenden, also als Fiktion einer utopischen Zukunft verstanden werden muß, so kleben die Figuren der Fleißer an einer Sprache, die ihnen nichts nützt, weil sie nicht die ihre ist.
Weil Brechts Figuren so sprachgewandt sind, ist in seinen Stücken der Weg zur positiven Utopie, zur Revolution gangbar. Hätten die Arbeiter bei Siemens das Sprachniveau der Arbeiter Brechts, hätten wir eine revolutionäre Situation. Es ist die Ehrlichkeit der Fleißer, die ihre Figuren sprach- und perspektivelos bleiben läßt .... Das Wichtigste der Fleißerschen Stücke ist das Verständnis für die, "auf die es ankommt" (Horváth). Die Masse der Unterprivilegierten. Gerade das Theater muß deren Möglichkeiten des Sprechens verfolgen, und die Fleißer hat das als erste praktiziert .... Die Fleißer hat uns gezeigt, daß es Dummheit als allgemeine menschliche Schwäche gar nicht gibt, daß vielmehr ein von Macht- und Profitstreben gelenkter gesellschaftlicher Prozeß die einen "dumm" und die anderen "gescheit" braucht, und sie also so werden läßt, rücksichtslos und verbrecherisch.' (Franz Xaver Kroetz, 'Liegt die Dummheit auf der Hand?', *Süddeutsche Zeitung,* 20/21 November 1971)

(*a*) Wie unterscheidet sich laut Kroetz die Sprache der Fleißer von der Bertolt Brechts?
(*b*) Was findet er an der Fleißer 'ehrlich'? Suchen Sie Textstellen, die dies belegen.
(c) Wie führt Kroetz die Sprache 'der Unterprivilegierten' auf gesellschaftliche Zustände zurück?

Volksstück

'Ausgehend von konventionellen Hör- und Sehgewohnheiten des Publikums, von der scheinbar vertrauten und heiteren Atmosphäre des Dialektes, dem Hintergrund regionaler und sozialer Begrenzung, "einfachen" Problemen und Konflikten, einem überschaubaren Personal in einer kleinen Welt, wird dem Zuschauer das "Heimatliche" zur unheimlichen und bedrückenden Enge verfremdet. Was früher, d. h. vor der Fleißer und Horváth, das zumeist komische Volksstück zeigte – Alltägliches in einer familiären, sozialen und lokalen Umwelt, nach außen schützend abgeschirmt und mit positiver

Konfliktlösung –, ist hier fast ausweglose Wirklichkeit, die sich häufig in Sprachnot und Sprachlosigkeit artikuliert. Die dem Dialekt zurückgewonnene Literaturfähigkeit dient nicht der Unterhaltung, sondern kritischer Analyse ... Den Autoren des "neuen" Volksstücks ist gemeinsam, daß sie die konfliktreiche soziale Realität von "unten" und in modellhaften Konflikten thematisieren ....' (Hugo Aust *et al.*, *Volksstück. Vom Hanswurstspiel zum sozialen Drama der Gegenwart* (Munich, 1989), 318)

'Charakteristisch für das Genre des "realistischen Volksstückes" ist: 1. daß es vornehmlich von der Masse der "kleinen Leute" handelt, wobei freilich "Volk" als Gegenstand, wie die Besucherstatistik lehrt, keineswegs per definitionem "Volk" als Publikum impliziert; 2. daß es Menschen aus der Unter- und/oder Mittelschicht nicht illusionistisch-unterhaltend wie das Millowitsch- oder Ohnsorg-Theater, sondern – als "Volkstheater gegen den Strich" (Mennemeier) – kritisch-realistisch darstellt; 3. daß es im Gegensatz zur gehobenen Standardsprache die Umgangssprache aller Schattierungen vom Jargon bis hin zum Dialekt verwendet; 4. daß es sich – anders als das hermetische Theater der Avantgarde – um Gemeinverständlichkeit bemüht und 5. daß es in der Regel als Gegenwartsstück im Unterschied zu den diversen Formen des historischen Dramas auftritt.' (W. Buddecke/H. Fuhrmann, *Das deutsche Drama seit 1945: Kommentar zu einer Epoche* (Munich, 1981), 145)

(*a*) Umreißen Sie mit Ihren eigenen Worten die Hauptmerkmale des kritisch-realistischen Volksstücks, wie sie in diesen Definitionen identifiziert werden.
(*b*) Inwiefern lassen sich diese Kriterien auf *Pioniere* anwenden?
(*c*) Wie setzt sich laut diesen Autoren das kritische Volksstück (i) vom früheren Volksstück, (ii) vom modernen Unterhaltungstheater, und (iii) vom Avantgarde-Theater ab?

Dramatische Vergleichstexte

1. Ödön von Horváth, *Kasimir und Karoline* (1932)
Erstes Bild (Auszug)
(Karoline hat sich mit ihrem Freund Kasimir, der am Tag zuvor arbeitslos geworden ist, gestritten. Auf dem Münchener Oktoberfest lernt sie Schürzinger kennen.)
Karoline sieht [*Kasimir*] *nach; wendet sich dann langsam dem Eismann zu, kauft sich eine Portion und schleckt daran gedankenvoll.*
Schürzinger schleckt bereits die zweite Portion
KAROLINE Was schauens mich denn so blöd an?
SCHÜRZINGER Pardon! Ich habe an etwas ganz anderes gedacht.

KAROLINE Drum.

Stille.

SCHÜRZINGER Haben Sie auch zuvor den Zeppelin gesehen?

KAROLINE Ich habe doch keine zugewachsenen Augen.

Stille.

SCHÜRZINGER Der Zeppelin, der fliegt jetzt nach Oberammergau.

KAROLINE Ja und dann wird er noch einige Schleifen über uns beschreiben.

SCHÜRZINGER Waren das Fräulein schon einmal in Oberammergau?

KAROLINE Schon drei Mal.

SCHÜRZINGER Respekt!

Stille.

KAROLINE Aber die Oberammergauer sind auch keine Heiligen. Die Menschen sind halt überall schlechte Menschen.

SCHÜRZINGER Das darf man nicht sagen, Fräulein! Die Menschen sind weder gut noch böse. Allerdings werden sie durch unser heutiges wirtschaftliches System gezwungen, egoistischer zu sein, als sie es eigentlich wären, da sie doch schließlich vegetieren müssen. Verstehens mich?

KAROLINE Nein.

SCHÜRZINGER Sie werden mich schon gleich verstehen. Nehmen wir an, Sie lieben einen Mann. Und nehmen wir weiter an, dieser Mann wird nun arbeitslos. Dann läßt die Liebe nach, und zwar automatisch.

KAROLINE Also das glaub ich nicht!

SCHÜRZINGER Bestimmt!

KAROLINE Oh nein! Wenn es dem Manne schlecht geht, dann hängt das wertvolle Weib nur noch intensiver an ihm – könnt ich mir schon vorstellen.

SCHÜRZINGER Ich nicht.

Stille.

KAROLINE Können Sie handlesen?

SCHÜRZINGER Nein.

KAROLINE Was sind denn der Herr eigentlich von Beruf?

SCHÜRZINGER Raten Sie doch mal.

KAROLINE Feinmechaniker?

SCHÜRZINGER Nein. Zuschneider.

KAROLINE Also das hätt ich jetzt nicht gedacht!

SCHÜRZINGER Warum denn nicht?

KAROLINE Weil ich die Zuschneider nicht mag. Alle Zuschneider bilden sich gleich soviel ein.

Stille.

SCHÜRZINGER Bei mir ist das eine Ausnahme. Ich hab mich mal mit dem Schicksalsproblem beschäftigt.

KAROLINE Essen Sie auch gern Eis?

SCHÜRZINGER Meine einzige Leidenschaft.
KAROLINE Die einzige?
SCHÜRZINGER Ja.
KAROLINE Schad!
SCHÜRZINGER Wieso?
KAROLINE Ich meine, da fehlt Ihnen doch was.
(Ödön von Horváth, *Kasimir und Karoline* (*Gesammelte Werke*, 5) Frankurt/ Main, 1986, 13-15)

2. Martin Sperr, *Jagdszenen aus Niederbayern* (1965)
Sechste Szene (Auszug)
(Tonka fühlt sich von ihrem Freund Abram verstoßen. Um sich zu rächen, bietet sie sich Volker, dem Liebhaber Marias, an.)

TONKA Schau mich an. Schau, wie jung ich bin.
VOLKER Du riechst nach Bier.
TONKA Ja. Aber jung bin ich! Wenn du zahlen kannst, bin ich zu haben. In Zukunft und überhaupt. Willst du mich?
Volker trinkt.
Sie hat dir doch verboten, daß du trinkst!
VOLKER Ja.
Sie umarmt ihn. Sie läßt ihn los. Er läuft zum Bier und trinkt.
TONKA Schmeckts nicht?
Er schüttelt den Kopf.
Wirfs weg!
Er wirft es weg.
Damit du die Händ frei hast. Jetzt hast du den gleichen Geschmack im Mund wie ich.
VOLKER Man kann sich ja gar nicht gegen dich wehren.
TONKA Du mußt zahlen.
VOLKER *greift nach ihr*: Ich will doch zahlen. Ich zahl ja.
Was kann ich mehr tun als zahlen. Warum machst du das alles mit mir! Und wenns der Abram erfährt, dann erschlag ich ihn.
Tonka klammert sich an ihn.
TONKA Er soll es erfahren! Wenn er mich so behandelt, dann mach ich das auch so. Grad extra! *Sie küßt ihn, läßt aber plötzlich los.* Nein. Ich darf nicht. Ich darf nicht. – Ich hab zuviel Bier.
Volker greift nach ihr.
VOLKER Was soll ich denn tun? Komm, renn nicht weg! Siehst doch, daß ich alles tu, was du willst.
TONKA Ich will – ja – aber – ich will nicht. Was du für Augen hast.
VOLKER Stell dich nicht an.

TONKA *hat Angst*: Schau mich nicht so an. Schau mich nicht so an! Ich hab zuviel Bier.

VOLKER Ich hab nicht lang Zeit. Ich zahl dann schon.

TONKA Also gut. *Sie beginnt, sich auszuziehen.*

VOLKER Du brauchst dich nicht auszuziehen. Es geht so.Sehen will ich dich gar nicht.

(Martin Sperr, *Bayrische Trilogie* (Frankfurt/Main, 1972), 31-32.

3. Franz Xaver Kroetz, *Mensch Meie*r (1977)
Erster Akt, 2. Szene (Auszug).
(Otto, seine Frau Martha und sein Sohn Ludwig schauen sich die Hochzeit vom König Karl Gustav IV von Schweden und der Deutschen Silvia Sommerlath im Fernsehen an).

MARTHA Was Rang und Namen hat auf der Welt, is da!

LUDWIG Fad is.

MARTHA Weilst kein Gefühl hast für die Schönheit. Wieviele Könige gibts denn noch auf der Welt? Die kann man an die Finger von einer Hand herunterzähln. Also is man dankbar, wenn man es erlebn kann, so ein Ereignis.

OTTO *hochdeutsch*: Epochemachend. Obwohl nix mehr zum Sagn ham, die "gekrönten Häupter".

MARTHA Sie soll sehr intelligent sein, die Silvia.

OTTO Aber er ned, da hat er recht. *Meint Ludwig.* Trotzdem: Bring es du einmal zu sowas wie der und zu seim Geld, dann kannst redn. Vorher nicht.

MARTHA Seids still, wo sie jetzt Mann und Frau gewordn sind. *Lacht.* Sie blinzelt ihm immer zu, siehst?

OTTO *lacht* Mistviech! Die hat ihn in der Hand. "Schau, schau, jetzt kannst nimmer aus, Männlein!" denkt sie sich.

MARTHA *muß auch lachen*: Ihr Männer!

OTTO Zu, die Falle! *Lacht, nickt.*

Pause.

MARTHA Was ist Glück? Da hörst es: sich selbst vergessn und für die andern da sein.

OTTO Die ham leicht reden.

MARTHA Was das Schöne is und mir gfallt, daß es eine echte Liebesheirat is! Er hats gesehn bei der Olympiade und gsagt: Die nimm ich, egal wer es is. – Das is wirklich königlich!

OTTO *lacht.*

LUDWIG Hunger!

MARTHA Eh schon alles fertig.

(Franz Xaver Kroetz, *Mensch Meier/ Der stramme Max/ Wer durchs Laub geht* ... (Frankfurt/Main, 1979), 11-12)

(*a*) Welche Gemeinsamkeiten in der Dialogführung lassen sich zwischen diesen Texten und *Pioniere* feststellen? Wie verhalten sich die Figuren einander gegenüber?
(*b*) Identifizieren Sie etwaige thematische Parallelen zu Marieluise Fleißers Text. Erinnern diese Passagen an bestimmte Stellen in *Pioniere*?
(*c*) Woran erkennt man das Volksstückhafte dieser Auszüge?

Pioniere in Ingolstadt im Spiegel der Kritik

'Marieluise Fleißer schildert die intimsten zwischenmenschlichen Beziehungen, die zwischen Mann und Frau. Sie illustriert durch ihr Stück, daß selbst die privatesten menschlichen Beziehungen geprägt sind durch das gesellschaftliche Sein der Menschen, und sie demonstriert, daß die gesellschaftlichen Zustände die zwischenmenschlichen Beziehungen brutalisiert und reduziert haben und daß nur der einigermaßen konfliktfrei leben kann, der sich diese Einsicht zu eigen macht und sich den Zuständen anpaßt .... Sie stellt das zwar ohne Aggression, aber voller Leidenschaft und mit aller Eindringlichkeit dar. Sie ist deshalb auch nicht eigentlich gesellschaftskritisch, weil ihre Intention nicht in erster Linie auf die Veränderung der von ihr geschilderten gesellschaftlichen Zustände hinausläuft. Sie zeigt den gesellschaftlichen Ist-Zustand, dem sie sich selbst auch hilflos ausgeliefert sieht. Die Vision vom anderen, besseren Leben scheint zwar in den Äußerungen der Gestalten im Stück immer wieder auf. Aber sie wird als vager hilfloser Appell benutzt, vor allem in Situationen besonders unerträglichen Drucks .... Oder sie erscheint als Lebenslüge und Selbstbetrug in den Äußerungen von Gestalten, die ... ganz am unteren Ende der gesellschaftlichen Skala rangieren.' (Müller, 85-8)

(*a*) Wie charakterisiert Müller Marieluise Fleißers Schilderung der zwischenmenschlichen Beziehungen im Stück? Trifft seine Auffassung Ihrer Meinung nach zu?
(*b*) Warum ist die Fleißer laut Müller nicht in erster Linie gesellschaftskritisch?
(*c*) Wo identifiziert Müller Visionen eines besseren Lebens im Text? Suchen Sie Beispiele hierfür.

'Die Sozialkritik der *Pioniere in Ingolstadt* bleibt aktuell, was nicht gleichbedeutend damit ist, daß die im Stück dargestellten Probleme unserem Alltag bis ins konkrete Detail entsprechen müssen .... Die politische Aussage der *Pioniere in Ingolstadt* bleibt so lange aktuell, wie unsere Gesellschaft strukturelle Ähnlichkeiten mit der autoritären Sozialordnung der Weimarer

Republik hat. Aber das Stück ist kein Agitationstheater; es hat keine direkte Aktualität in dem Sinne, daß die dargestellten Probleme unmittelbar als die unseren erkennbar wären. Einen Skandal wird das Stück, obwohl die Neufassung stärker politisch akzentuiert ist, nicht mehr hervorrufen. Was nicht heißt, daß die *Pioniere in Ingolstadt* politisch wirkungslos geworden seien.' (Kässens/Töteberg, '"... fast schon ein Auftrag von Brecht"', 116/17)

(*a*) Was verstehen Sie unter einer 'autoritären Sozialordnung'? Worin manifestiert sich eine solche (i) im Stück und (ii) in der heutigen Welt?

'Betrachtet man die Themenbereiche, die Fleißer aufgreift, im Vergleich mit dem Themenspektrum der feministisch orientierten Frauenliteratur, so läßt sich eine gewisse Vollständigkeit feststellen, wenn auch die Gewichtung und Darstellung notwendigerweise anders ausfallen müssen. Die Kritik an der patriarchalischen Gesellschaft wird nicht offen und konkret angesprochen oder alternativen Lösungsvorschlägen gegenübergestellt. Es zeigt sich insofern eine emanzipatorische Tendenz, als Fleißer als "Chronistin patriarchaler Sitten und Gebräuche" ein Bild der Gesellschaft festhält, das in keinem Fall eine "heile Welt" repräsentiert. Bei den einzelnen Frauengestalten – egal wie weit sie auf ihrem Weg zur Emanzipation fortgeschritten sind – wird jeweils deutlich, daß es sich immer um einen Entwicklungsbogen handelt, in dem die Figur bewußt oder unbewußt versucht, ihre Wirklichkeit bzw. ihre Identität zu finden. Dieser Aspekt ist in der heutigen Frauenliteratur ... in ähnlicher Weise zu finden .... Um dem Anspruch feministischer Ästhetik zu genügen, müssen Momente des bewußten Widerstands, Protest oder sogar Kampfansage vorhanden sein, wie das bei vielen Autorinnen der Frauenbewegung der Fall ist. Einige Elemente sind bei Marieluise Fleißer durchaus vorhanden. Als Nicht-Feministin schreibt sie in zunehmendem Maße mit dem kritischen Bewußtsein ihrer Geschlechtszugehörigkeit, was die Entwicklung der Frauencharaktere beweist.' (Preuß, 191/2)

(*a*) Wo sieht Patricia Preuß die 'emanzipatorische Tendenz' der Fleißer?
(*b*) Warum wird Marieluise Fleißer als 'Nicht-Feministin' bezeichnet?
(*c*) Inwiefern läßt sich diese Definition der 'feministisch orientierten Frauenliteratur' auf *Pioniere* anwenden?
'Das Werk der Fleißer – so eng es auch zu sein scheint – legitimiert sich durch seine Sprache. Bevor man auf die Figuren und Inhalte trifft, nimmt man die merkwürdige Fremdheit dieser Sprache wie etwas doch Vertrautes auf. Da es der bayerische Dialekt nicht ist ... half man sich in den Beschreibungen immer mit dem Begriff "bayerische Diktion". Damit waren Tonfärbung und Tonfall, auch Eigentümlichkeiten der Wortwahl gemeint, die in den hochdeutschen Satz immer eine volkstümliche Fremdheit importieren ....

Alles, was man bei der Fleißer an Sprachentwicklung wahrnimmt, ist keinesfalls als ein bewußtes Suchen nach einer neuen, ausdruckskräftigen und haftenden Sprache zu verstehen. Darin unterschied sie sich sicher von Brecht. Ihre Sprache hat von Anfang an eine hohe Verdichtungskraft, eine starke Innenspannung und Energie. Jeder Satz ist so, als müsse sie eine neue Wand durchstoßen .... Die Sätze erscheinen in ihrer konzentrierten Knappheit als kalkuliert; aber sie sind nur herausgedrückt aus einer Psyche, die mit ihrem Stoff aggressiv verbunden ist.' (Rühle, 'Leben und Schreiben', 39-41)

(*a*) Inwieweit läßt sich Pioniere als Beispiel für Marieluise Fleißers 'enges' Werk anführen?
(*b*) Wie charakterisiert Rühle die Sprache der Fleißer? Suchen Sie Textstellen, die seine Thesen bestätigen/ widerlegen.

Select bibliography

Primary literature

Marieluise Fleißer, *Gesammelte Werke*, ed. Günther Rühle, 3 volumes (Frankfurt, 1972); volume 4, ed. Günther Rühle and Eva Pfister (Frankfurt, 1989).

Secondary literature

Arnold, Heinz Ludwig (ed.), *Text und Kritik: Marieluise Fleißer*, 64 (Göttingen, 1979). Collection of essays on aspects of Fleißer's life and work.

Aust, Hugo *et al.*, *Volksstück: Vom Hanswurstspiel zum sozialen Drama der Gegenwart* (Munich, 1989). Detailed history of the genre: useful for the literary–historical context of Fleißer's plays, to which a section is devoted.

Cocalis, Susan L., 'Weib ohne Wirklichkeit, Welt ohne Weiblichkeit. Zum Selbst-, Frauen- und Gesellschaftsbild im Frühwerk Marieluise Fleißers', in von der Lühe, Irmela (ed.), *Entwürfe von Frauen in der Literatur des 20. Jahrhunderts* (Berlin, 1982), 64-85. Somewhat overstated analysis of the position of women in Fleißer's early work. Sees females in search of stability in a crumbling patriarchal world, a search which produces ever greater entrapment.

Hoffmeister, Donna L., *The Theater of Confinement – Language and Survival in the Milieu Plays of Marieluise Fleißer and Franz Xaver Kroetz* (Columbia, South Carolina, 1983). Excellent analysis of dialogue in the play, showing in some detail how it enacts a total lack of communication between characters.

Karasek, Hellmuth, 'Zur Erneuerung des Volksstücks: Auf den Spuren Marieluise Fleißers und Ödön von Horváths', in Arnold, H. L. and Buck, T. (eds.), *Positionen des Dramas. Analysen und Theorien zur deutschen Gegenwartsliteratur* (Munich, 1977), 137-69. Survey of the critical 'Volksstück' of the 1960s and 1970s, as influenced by Fleißer and Horváth.

Kässens, Wend and Töteberg, Michael, '" ... fast schon ein Auftrag von Brecht". Marieluise Fleißers Drama *Pioniere in Ingolstadt*', in Fuegi, J. *et al.* (eds.), *Brecht-Jahrbuch 1976* (Frankfurt/Main, 1976), 101-19. Assesses the play as an early exercise in epic theatre and relates it to Brecht's ideas on gestic language.

—, *Marieluise Fleißer* (Munich, 1979). Good introduction to Fleißer's life and plays, with useful details of the works in performance.

Kord, Susanne, 'Fading Out: Invisible Women in Marieluise Fleißer's Early Dramas', *Women in German Yearbook*, 5 (1989), 57-72. Explores 'how Fleißer's dependence on her male mentors finds expression in her early dramas', with rather predictable results: in *Pioniere*, self-definition and identification are seen as impossible.

Kraft, Friedrich (ed.), *Marieluise Fleißer: Anmerkungen, Texte, Dokumente* (Ingolstadt, 1981). Collection of essays and documents on Fleißer, with valuable contributions by Eva Pfister and Günther Rühle on her life and work.

McGowan, Moray, *Marieluise Fleißer* (Munich, 1987). By far the best general study of Fleißer's life and work to date. Thoroughly researched, combining information with lucid interpretation. Good chapter on *Pioniere*.

Macher, Hannes S., '"Ich schreibe Leben – aus Betroffenheit". Marieluise Fleißer: *Pioniere in Ingolstadt*', in Weber, Albrecht (ed.), *Handbuch der Literatur in Bayern* (Regensburg, 1986), 539-49. Brief but useful introduction to the play, its background and reception.

Mennemeier, F. N., *Modernes deutsches Drama 2* (Munich, 1975). Has a challenging section on 'Volkstheater gegen den Strich', including Fleißer, which laments the lack of a clear critical perspective in the plays.

Müller, Gerd, *Das Volksstück von Raimund bis Kroetz* (Munich, 1979). Survey of the development of the genre, containing an excellent analysis of *Pioniere*.

Preuß, Patricia, '"Ich war nicht erzogen, daß ich mich wehrte". Marieluise Fleißer und ihr Werk in der Diskussion um weibliches Schreiben', *Germanic Review*, 62 (1987), 186-93. Sober assessment of the works in terms of their feminist impulse. Fleißer seen as precursor of later, more overtly feminist writing.

Rühle, Günther (ed.), *Materialien zum Leben und Schreiben der Marieluise Fleißer* (Frankfurt, 1973). Invaluable compilation of essays, reviews and other documents on and by Fleißer. Of particular value for *Pioniere* are the reviews of the various productions and a series of interviews Fleißer gave during the 1960s and 1970s,

—, 'Leben und Schreiben der Marieluise Fleißer aus Ingolstadt', in I, 7-60. Pioneering analysis of Fleißer's achievement. Informative, sensitive, and elegant.

—, 'Rückblick auf das Leben und Schreiben der Marieluise Fleißer', in IV,

549-70. Excellent assessment of Fleißer's achievement in the light of recent developments and newly published texts.
Stritzke, Barbara, *Marieluise Fleißer: Pioniere in Ingolstadt* (Frankfurt, 1982). Only full-length study of the play, covering its background, genesis, reception,themes and structure. Rather fragmentary.
Tax, Sissi, *Marieluise Fleißer: Schreiben, Überleben. Ein biographischer Versuch* (Basel/Frankfurt, 1984). Detailed biography of the years 1901-38, along with photographs, letters and documents. Strong feminist slant.
Töteberg, Michael, 'Die Urfassung von Marieluise Fleißers *Pioniere in Ingolstadt*', *Maske und Kothurn*, 23 (1977), 119-21. Short review of the early versions of the text.

Vocabulary

The vocabulary covers all three main sections of this edition, i.e. Introduction, *Pioniere in Ingolstadt* and *Arbeitsteil*. Words are given only in the meaning in which they appear in this text, and plurals are indicated only where relevant. Vowel changes in irregular verbs are given in abbreviated form. The abbreviations used are: coll (colloquial), jur (legal), mil (military), photo (photographic), S. Ger (South German regional usage), sl (slang).

der **Aal** eel
ab.fahren lassen, ie, a (jdn) (sl) to send packing
abgedankt retired
abgekämpft worn out
abgeneigt averse
abgeschlagen exhausted
abhängig dependent
das **Abhängigkeitsverhältnis, -se** dependency relation
ab.lassen, ie, a to stop
ab.legen to give up
ab.nehmen (jdm etw) to believe
ab.saufen, o, o (coarse) to drown
ab.schirmen to shield
sich **ab.setzen von** to contrast with
ab.sprechen, a, o to agree
ab.spülen to wash up
ab.streiten, i, i to deny
ab.stürzen to fall
Abtreten! (mil) dismissed!
ab.tun, a, a (jdm etw) (sl) to put, run sb down
ab.verlangen (jdm etw) to demand of
die **Abwechslung** change, distraction
sich **ab.wenden, a, a** to turn away
der **Abzug** withdrawal
die **Achseln** (pl)(mil) stripes, epaulets
die **Achseln schupfen** to shrug one's shoulders
der **Acker, Ä** field
ahnen to suspect
die **Ähnlichkeit** similarity
akkurat precisely
die **Aktualität** topicality, relevance
aktuell topical
allerdings of course, admittedly
allerhand a great deal
die **Altane** terrace, balcony
die **Ameise, -n** ant
sich **an.bieten, o, o** to offer oneself
das **Andenken** souvenir
an.führen (als Beispiel –) to give as an example
die **Angabe** statement
angeblich allegedly
an.gehen, i, a to concern
angelehnt leaning
der **Angelpunk**t pivotal point, anchor
die **Angemessenheit** appropriateness
angetrunken inebriated
angewiesen sein (auf) to have to make do with
an.greifen, i, i to attack
der **Angriff** attack
der **Anhalt** grounds
an.hängen (jdm etw) (coll) to pin sth on sb
sich **an.hängen** (coll) to become attached
die **Ankerwinde** capstan
sich **an.lachen (jdn)** (coll) to pick sb up

der Anlaß cause
anläßlich (+ gen) on the occasion of
sich an.passen (+ dat) to adjust, adapt to
die Anregung suggestion
an.reichern to enrich
anscheinend apparently
an.scheißen, i, i (coarse) to do the dirty on sb
der Anschlag attack
sich anschmeißen, i, i (an jdn) (coll) to be all over sb
an.singen, a, u to make up to sb, trick sb
eine Ansprache haben to have contact with others
an.sprechen, a, o to mention, touch upon
der Anspruch requirement, claim
Ansprüche stellen to have high standards
anspruchsvoll sophisticated
der Anstand decency
anständig decent
an.stellen (etw) to be up to sth
sich anstellen (coll) to make a fuss
einen Antrag machen (jdm) to propose (marriage)
an.treten, a, e (mil) to fall in
an.wenden (auf) to apply to
das Anzeichen, - indication, sign
an.zeigen to report to the police
der Anzeigenteil advertisement section
der Apparat camera
das Arbeitsamt employment office
das Arbeitsgeräusch, -e sound of working
arbeitslos unemployed
arg (das ist mir -) that's awful
der Arm (am langen - sitzen) to have influence
armselig wretched, miserable
der Arsch (jdn am - lecken) (coarse) to kiss sb's arse, fuck off
das Arschloch, -¨er (coarse) arse-hole
der Aufbruch departure
auf.drängen (jdm etw) to force sth on sb
auf einmal all at once, suddenly
auf.fallen, ie, a to be conspicuous
auffällig conspicuous
die Auffassung, -en view, conception
die Aufführung production, performance
die Aufgabe task
aufgehoben (gut -) safe
aufgeputzt (coll) dressed up
der Aufgeschlossene, -n open-minded person
auf.greifen, i, i to take, pick up
auf.keimen to dawn
auf.kommen, a, o to get out
die Auflage version
der Auflauf, äe crowd
der Aufpasser (den - machen) (coll) to be on the lookout
die Aufregung excitement
auf.scheinen, ie, ie to appear
die Aufschrift sign
auf.sitzen, a, e (jdm) to harass, pester
der Auftrag commission, instruction
auf.treten, a, e to appear, occur
auf.trumpfen to crow
auf.ziehen, o, o to tease; to raise (children)
der Augiasstall Augean stables, pigsty
aus.baden (coll) to carry the can
die Ausbeutung exploitation
die Ausbildung training
sich aus.drücken to express oneself
ausdruckskräftig expressive
aus.fallen, ie, a to turn out
aus.fliegen, o, o to go out
aus.fressen, a, e (coll) to suffer for
Ausgang haben/nehmen (mil) to be off duty, have/take time off
ausgehend von taking as its starting point
ausgeliefert (+dat) at the mercy of
ausgesprochen marked
das Ausgleiten slip
sich aus.kennen, a, a to know one's way around, know what's going on

aus.lassen, ie, a (etw an jdm -) to vent sth on sb
sich **aus.lassen, ie, a** to let off steam
aus.machen to settle, agree
die **Ausnahme** exception
aus.pfeifen, i, i to boo
aus.richten to pass on (a message)
aus.rotten to eradicate, wipe out
die **Ausrüstung** equipment
aus.sagen to say, state
die **Aussage** statement, message
ausschlaggebend decisive
ausschweifend wild
die **Äußerung, -en** statement
aus.sehen, a, e (nach etw) to look like
die **Aussicht** chance
aus.sprechen, a, o to express
aus.stellen to display, exhibit
aus.treiben, ie, ie (jdm etw) to cure sb of sth
die **Auswirkung, -en** effect
sich **aus.ziehen, o, o** to undress
ausweglos inescapable, hopeless
aus.weichen, i, i to go elswewhere
aus.wischen (jdm eins -) (coll) to get one over on sb
der **Auszug, üe** extract

die **Bahn (in der -)** on the right lines
der **Bahnhof (keinen - kennen)** not to have any scruples
der **Balg** (sl) brat
der **Balken** beam
die **Bande** gang
der **Barras** military
basteln to build
baufällig dilapidated
die **Baustelle** building site
der **Bautrupp** construction squad
der **Bazillus, -en** germ
beabsichtigen to intend
die **Bearbeitung** adaptation
beben to tremble
sich **bedecken** to cover oneself up
das **Bedenken, -** reservation
die **Bedeutung** significance
die **Bedienung** service
die **Bedingtheit** conditioning, dependence
die **Bedingung** condition
bedrohen to threaten
bedrückend oppressive
das **Bedürfnis** need
die **Beeinträchtigung** diminution
die **Beengtheit** crampedness
sich **befassen (mit)** to do sth about sth
zu **Befehl (mil)** Sir!
die **Beförderung** promotion
der **Befreiungsversuch, -e** attempt to break free
befriedigen to satisfy
die **Begeisterung** enthusiasm
die **Begrenzung** restriction
der **Begriff** concept
bei.biegen, o, o (coll) to find
bei.kommen, a, o (+ dat) to cope with, deal with
beisammen together
sich **beklagen** to complain
belegen to support, substantiate
beleidigt offended
sich **bemühen (um etw)** to strive for, aim to gain
benachteiligen to treat unfairly
das **Benehmen** behaviour
sich **benehmen, a, o** to behave
die **Beobachtung** observation
berechtigen to justify
die **Bereitschaft** readiness
sich **besaufen, o, o** (sl) to get drunk
sich **beschäftigen (mit etw)** to concern oneself with sth
bescheißen, i, i (coarse) to cheat
beschweren to weigh down
sich **beschweren** to complain
die **Beschwörung** entreaty
sich **besinnen, a, o** to think
bestehen, a, a (in) to consist in, lie in
bestimmen to dominate, determine
die **Bestimmte** girlfriend
die **Bestimmung, -en** requirement
bestrafen to punish
bestürzt dismayed, filled with consternation
die **Betroffenheit** consternation

sich beugen to bend down
die Bevölkerung population
bewachen to guard
bewußt conscious
bezeichnen (mit) to put, estimate at ; **(als)** to describe as
die Beziehung, -en relationship
der Bezirk domain
in bezug auf with regard to
bezweifeln to doubt
das Bierzelt beer tent
sich bilden to form
blamieren to show up
blank polished, shining
das Blatt (sich kein - vor den Mund nehmen) not to mince one's words
die Blattern (pl) pock marks
das Blickfeld focus, view
blöd stupid
bloß.legen to expose
blütenweiß snow white
das Bollwerk bastion, stronghold
die Branche trade
brandmarken to brand
brave good, decent
die Brauerei brewery
die Braut, -¨e fiancée
das Brückenbauskelett scaffolding
sich bücken to bend down
der Büffel (coll) idiot
die Bühne stage
der Bühnenwitz stage joke
büßen to pay for, atone for

danach (jdm ist -) to be in the mood
darauf.kommen, a, o (jdm) (coll) to catch
die Darbietung presentation
die Darstellung portrayal
der Daumen (den - auf jdn drücken) to keep sb under control
der Deckel lid
die Denkweise, -n attitude
derb coarse
dick.haben, a, a (sl) to be fed up with
im Dienst on duty
der Dienstgrad, -e (mil) higher ranks
das Dirnentum prostitution
dran.kommen, a, o to have one's turn
drauf.drücken to harass, put pressure on
der Dreh the knack, hang (of something)
dringend urgent
drohen to threaten
der Druck pressure
durchgehen lassen, ie, a (jdm etw) to let sb get away with sth
der Durchschlag carbon copy
sich durch.setzen to assert oneself
durchstoßen, ie, o to break through

das Ehrenwort word of honour
die Ehrlichkeit honesty
die eidesstattliche Erklärung affidavit
eifersüchtig jealous
eigen (sich zu- machen) to adopt
die Eigenart, -en characteristic
im Eigenbau home-made
eigenmächtig werden to act without permission
die Eigentümlichkeit, -en peculiarity
ein.biegen, o, o to turn into
sich ein.bilden to imagine
der Einblick insight
die Eindringlichkeit forcefulness
eingebaut organised, arranged
das Eingesperrtsein imprisonment
der Einhalt (- gebieten) to stop
sich ein.halten (an jdn) (S. Ger) to hang, hold on to sb
eingekeilt wedged in
die Einheit (mil) unit
einigermaßen to an extent
sich ein.krallen (coll) to get down to it
die Einkunft, -¨e income
ein.räumen to grant
sich ein.reihen to get into position
ein.reißen, i, i (coll) to become a habit
ein.schätzen to assess
der Einschlag element

ein.schnaufen (jdn - wie Luft) to treat sb as if they didn't exist
ein.sehen, a, e to see into; to understand
die **Einstellung** attitude
ein.weihen to open officially
der **Eintritt** admission fee
das **Einzelfoto, -s** individual photograph
einzigartig unique
die **Empörung** outrage
endgültig final
energisch firm, forceful
die **Enge** restriction; **(in die - treiben)** to force into a corner
die **Entfremdung** alienation
sich **entgehen lassen, ie, a** to miss
enthüllen to reveal
entlassen, ie, a (mil) to dismiss
entmilitarisieren to demilitarise
entschlossen determined
entschlüpfen to escape from
entsprechen, a, o to appeal to; correspond with
entstehen, a, a to result
enttäuschen to disappoint
entwickeln to develop
der **Entwicklungsbogen** process of development
die **Erbin** heiress
der **Erdboden** earth
das **Ereignis** event
erfahren, u, a to find out
das **Ergebnis** result
erkennen, a, a to recognise
erläutern to explain
erleben to experience
erpressen to blackmail
ersaufen, o, o (sl) to drown
erschrecken, a, o to be startled, frightened
ersetzen to replace
ertrinken, a, u to drown
sich **erweisen, ie, ie (als)** to prove
die **Eskalierwand** scaling wall
etwaig any, possible
extra on purpose

fad(e) boring
die **Falle** trap
fälschen to falsify, distort
faßbar comprehensible
faulen to rot
fehlen to be missing
Feierabend machen to finish work, have time off
feig cowardly
die **Feindschaft** hostility
die **Feldesse** (mil) field stove
der **Feldwebel** (mil) sergeant
fest.stellen to find
die **Festung** fortress
filzen (coll) to search
die **Flamme** (coll) girlfriend
vom **Fleck weg** on the spot
fliegen, o, o (auf jdn) (coll) to fall for sb
fluchen to curse
der **Fluchtweg** escape route
foppen to make fun of, mock
fordern to demand
fort.schreiten, i, i to progress
die **Frauenbewegung** women's movement
der **Frauenüberschuß** surplus of women
Freiübungen (pl) **machen** to do one's exercises
freiwillig voluntary
die **Fremdheit** strangeness
fügen to connect, join
funkeln to sparkle
der **Funke** spark
die **Fußangel, -n** mantrap

eine **galante Krankheit** sexual disease
gangbar possible
die **Garnisonsstadt** garrison town
geben, a, e (es jdm -) (coll) to show sb, put sb in their place
der **Gebrauch, -¨e** custom
gefaßt (auf) prepared for
der **Gefreite** private
die **Gegenreformation** Counter Reformation
der **Gegenstand** subject
gegenüber.stellen to confront
das **Gegenwartsstück** play set in the

present
der **Gegenwert** one's money's worth
gehemmt inhibited
gehen, i, a (nach jdm -) to be up to
gehoben elevated
gehorchen to obey
der **Gehsteig** pavement
die **Geilheit** lecherousness
die **Geißel** scourge
das **Gelände** area
der **Gelernte** expert, dab hand
gell (S. Ger, interj) isn't it? right?
geltend machen to assert
gemein mean, rotten
der **Gemeine** (mil) common soldier
die **Gemeinheit** meanness, nastiness
die **Gemeinsamkeit, -en** common features
die **Gemeinverständlichkeit** accessibility
das **Gemüt** soul
geraten, ie, a to slip
gerecht just, fair
gescheit clever
das **Geschirr** crockery
geschlagen defeated
das **Geschlecht** sex
der **Geschlechtsakt** sexual intercourse
die **Geschlechtszugehörigkeit** gender
der **Geschmack** taste
gesellschaftsimmanent inherent in society
gesetzlich legal, legitimate
die **Gesinnung** beliefs
gespannt taut
die **Gestalt, -en** character
gestalten to depict, present
das **Gestell** frame, body
das **Gewand** dress
gewandt competent, clever
die **Gewichtung** evaluation
das **Gewissen** conscience
gewitzt crafty
die **Gewohnheit, -en** habit
die **Glacisbank** bench on a glacis (slope)
glatt simply, just
gleichbedeutend synonymous
gleichgeschaltet forced into the (Nazi) party line
im **Gleichschritt** (mil) in step
glotzen (coll) to stare
die **Gnade** grace, mercy
Gnade Gott God help
gnädig graciously, beautifully
der **Grad** level
der **Grimm** wrath
der **Großhandel** wholesale business
die **Grube** pit, hole
grundlegend fundamental
grüßen (mil) to salute
die **Gültigkeit** validity
gut (jdm - sein) to like, love sb

haftend memorable
das **Hakenkreuz** swastika
handgreiflich violent
der **Handgriff, -e** adjustment, touch
hand.lesen, a ,e to read palms
die **Handlung** plot
hängen lassen (jdn) (coll) to let sb down, leave sb in the lurch
hantieren (an) to fiddle about with
hartnäckig stubborn
der **Häuptling** chief, boss
das **Hauptmerkmal, -e** main feature
die **Haut** skin
heben, o, o to lift, raise, elevate
die **Hecke** hedge
die **Heeresvorschrift, -en** army regulation
der **Hehl (keinen - machen)** to make no secret
heilen to cure
der **Heilige, -n** saint
heimtückisch malicious
der **Heiratsantrag** marriage proposal
heiter cheerful
heran.können, o, o (an jdn) to get to sb
sich **heran.machen** (coll) to approach, chat up
sich **heran.pirschen** (coll) to creep up
heran.schlendern to stroll up
heraus.bringen, a, a to find out
sich **heraus.halten, ie, a** to keep out of it

heraus.rücken (coll) to cough up
heraus.schinden (coll) to squeeze out
her.gehen, i, a (mit jdm) to go out with sb
her.langen to feel
die **Herrschenden** (pl) the ruling classes
sich **her.stellen** (S. Ger) to give it a try
herrschen to reign
hervor.rufen, ie, u to cause
herum.schwirren to buzz around
herum.tun, a, a (mit jdm) to mess about with
hetzen to be in a rush; to hurry sb
heucheln to feign
die **Heuchelei** hypocrisy
die **Hilfskraft, -¨e** assistant
hinaus.laufen, ie, au (auf) to aim at
hinaus.schmecken to gain experience, see the world
hinein.stecken to invest in, do for
hinein.legen to trick
sich **hinein.reiten, i, i (in etw)** to get into trouble, let oneself in for sth
hinein.schlittern to slide into
hin.fassen to grab hold of
die **Hingabe** devotion
hin.hängen (jdn) (coll) to rat on sb
sich **hin.hängen (an jdn)** (coll) to fall for, be after sb
hinken to limp
hinsichtlich (+ gen) with regard to
hinten und vorn nichts nothing at all
hinterfotzig underhanded
hinunter.schwemmen to drown
hinunter.stürzen to gulp down
hinzu.fügen to add
hinzu.kommen, a, o to be added
das **Hirn** brain
der **Hirsch, -en** (sl) bastard
hochdeutsch standard German
der **Hochstapler** con man
der **Holzsteg** wooden walkway, bridge
Hopp! (coll) up you get!
die **Horneule** (S. Ger) idiot
die **Hundertschaft, -en** group of one hundred
der **Hundling** cur

immerzu (here for 'immerhin') anyway
das **Imponiergehabe** display, attempt to impress
der **Informant** interviewee
der **Inhalt** contents
die **Innenspannung** inner tension
inspizieren to inspect
die **Innerlichkeit** inwardness
der **Intimverkehr** intimate relations
sich irren to be mistaken

jeweils in each case
johlen to howl
der **Josef** (coll) bloke, fellow
jüdisch Jewish
die **Jungfrau, -en** girl

in **Kabinettform** (photo) cabinet size
die **Kampfansage** declaration of war
die **Kanaille** (sl) cow, bag
kappen to cut
die **Kartei** records
die **Kasematte, -n** fortifications
die **Kaserne** barracks
kassieren to collect the money
kehren (mil) to about turn
kennzeichnen to characterise
kippen to tip
kitschig sentimental
klappen to rattle
klauen (coll) to steal, pinch
kleben (an) to be stuck with
das **Kleinbürgertum** lower middle class
das **Kloster** convent
knapp brief
die **Knappheit** conciseness
die **Kniebeuge (eine - machen)** to bend one's knees, genuflect
der **Kniff, -e** crease
kommen lassen, ie, a (nichts - auf jdn) (coll) not to want to hear a

bad word about sb
die **Konfliktlösung** resolution of conflict
kopfüber headfirst
der **Korb (jdm einen - geben)** (coll) to refuse sb
die **Kralle, -n** claw
der **Krampf** (coll) nonsense
die **Kränkung** insult
kriechen, o, o to crawl
das **Kriterium, -ien** criterion
der **Kumpel** (sl) mate, pimp
Kunststück! (coll) great!
den **kürzeren ziehen** to come off worst

der **Lahmarsch, äe** (sl) slowcoach
auf der Lauer liegen, a, e to lie in wait
läufig on heat
der **Laufschritt** (mil) quick march
laut (+ dat) according to
lauter sheer
der **Lebemann** man of the world
die **Lebenslüge** life-long illusion
der **Lebenswandel** (decent) way of life
ledig single, unmarried
legen (jdm etw) to set, lay
die **Lehne** backrest
die **Lehre** moral
die **Leiche** corpse
sich **leicht.machen** to loosen, remove clothing
leiden, i, i to suffer, bear
die **Leidenschaft** passion
die **Leier** scale, list
sich **leisten** to afford
die **Leiter** ladder
die **Lektion** lesson
der **Lieberdienerische** creep, crawler
der **Liebhaber** lover
lieblich charming, wonderful
die **Literaturfähigkeit** literary acceptability
locken to entice
lockern to loosen
locker (ein - Vogel) loose, bold person
die **Lockerung** slackening
das **Lokal** pub, bar
der **Lösungsvorschlag, -¨e** suggested solution
die **Lunge (- haben)** (coll) to have enough breath

der **Magen (keinen - haben)** (coll) not to be in the mood
sich **malen (etw)** to imagine
der **Männerturnverein** men's sports club
das **Mannsbild** (coll) bloke
die **Mannschaft** (mil) other ranks
die **Marie** (sl) money
das **Mark (durch - und Bein gehen)** to become second nature
eine(n) **markieren** (sl) to make a play for sb; to act big
das **Maß(in zunehmendem -)** increasingly
der **Massel** (coll) good luck
der **Maßkrug** one-litre tankard
der **Max** (coll) friend, fellow
die **Meldung** report, confession
mer (S. Ger) we
merkwürdig strange
messen, a, e to measure
der **Metzger (- spielen)** (coll) to go wild
der **Mißstand, -¨e** evil, shortcoming
das **Mistviech** (S. Ger, coarse) bitch
mit.liefern to provide
mit.teilen to communicate
der **Mitwisser** (jur) accessory
mobil fit
mogeln to cheat
morsch rotten, brittle
mustern to scrutinise

nach und nach gradually
die **Nachfrage haben** to be in demand
nach.geben, a, e to give in
nach.gehen, i, a (jdm) to follow
nach.holen to make up
nach.weinen (jdm) to shed tears over sb
nachweislich demonstrably
nachtragend unforgiving
die **Nachtschicht** night shift

der **Nacktspaß, -¨e** naked frolic
unter **den Nagel reißen, i, i** (coll) to steal
nageln to nail down
sich **nähern** to approach
zu **sich nehmen, a, o** to drink
der **Neid** envy
der **Nestbeschmutzer** one who fowls one's own nest
neuerdings recently
die **Neufassung** new version
die **Neuvermählte** newly-wed wife
not.tun, a, a to be necessary
die **Notwehr** self-defence
nüchtern sober

der **Oberfeld(webel)** (mil) staff sergeant
die **Oberfläche** surface
offenbaren to show, reveal
die **Ohnmacht** powerlessness
ohnmächtig impotent, powerless
ordinär vulgar

pachten (coll) to have a monopoly on
packen (coll) to manage
pappen (- bleiben) (coll) to get stuck
die **Parole** (mil) password
der **Passierschluß** (mil) end of evening leave
passen (es paßt) (coll) it's going well
das **Pech** bad luck
sich **pelzen** (coll) to loaf about
das **Personal** cast
Pfeif! (coll) Wow!
das **Pfeifkonzert** chorus of catcalls
pfeilgrad (coll) to be honest, in a word
Pfui (coll) ugh (expression of disgust)
das **Pissoir** urinal
der **Platz (fehl am -)** out of place
platzen to explode, burst
die **Positur (sich in - stellen)** to pose
die **Potenz** power, genius
prägen to dominate
präpariert treated
prellen to cheat, swindle
das **Pressieren** haste
der **Provinzmief** small-town atmosphere

quälen to torment
die **Quittung** receipt

sich **rächen** to avenge oneself
der **Rand** (dirty) rim
Rang und Namen haben to be somebody
rangieren to rank
sich **rar machen** (coll) to play hard to get
rasend mad, wild
die **Raubtierschaft** predatoriness
raus.hängen lassen (coll) to show it
rechtfertigen to justify
rechtschaffen decent, honest
die **Rede verschlagen, u, a (jdm)** to leave sb speechless
die **Regieanweisung** stage direction
reichen to be adequate
reif mature, ready
sich **reißen, i, i (um etw)** (coll) to be desperate for sth
reißend raging
reizen to attract
requirieren (mil) to requisition
reuen to move sb to pity
sich revanchieren to pay back, return the compliment
der **Rezensent** reviewer
sich **richten** to arrange, sort out, see to
sich **richten nach** to comply, fit in with
riechen, o, o to smell, sense
robben to crawl on one's stomach
sich **rollen** (coll) to be on one's way
der **Röntgenblick** x-ray sight
das **Roß** (coll) fool
die **Rotunde** rotunda
die **Routineuntersuchung** routine inspection
rücksichtslos ruthless
das **Rudelverhalten** pack mentality

das Rudergeräusch, -e sound of rowing
der Rüffel (einen - einstecken) (coll) to take a severe rebuke
sich rühren to do sth
ruppig rough

der Sachverhalt, -e fact
säen to sow
der Samen seed
die Sammeladresse joint address
sämtlich all
die Sau sow
der Saufbruder (sl) drinking partner
der Sausteg (sl) bloody walkway
die Schadenfreude gloating
schadenfroh gloating
sich schämen to be ashamed of oneself
die Schande disgrace
die Schattierung, -en shade
der Schaukelstuhl rocking chair
das Schauspringen diving exhibition
die Scheidung parting
scheinbar apparently
der Scheinwerfer spotlight
die Scheißbrücke (sl) bloody bridge
scheitern (an) to break down
die Scherbe, -n fragment
die Scheuklappenerziehung blinkered education
die Schicht class
das Schicksal fate
die Schikane, -n bullying, harassment
schikanieren to harass
schief.gehen, i, a to go wrong
schimpfen to scold
schinden, i, u to drive, flog
der Schinder slavedriver
Schißhaben (sl) to be scared
das Schlachtfeld battlefield
die Schlägerei, -en fight
das Schlagwort catch-phrase
der Schlamm mud
die Schlampe (coll) slut
schlau clever
schlauchen to wear out
schlecken to lick
die Schleife, -n loop
schlendern to stroll, amble
schleppen to drag, tow
schließen, o, o (aus etw) to conclude from
die Schlinge loop
schmachten to yearn
schmierig smutty
der Schneid (coll) guts
schnappen to catch, grab
schnaufen to breathe
schöpferisch creative
die Schraube, -n bolt, screw
der Schreiner carpenter
schubsen (coll) to nudge, push
der Schuft sneak, cad
schuften (coll) to work hard
der Schuldige guilty party
der Schutzmann policeman
die Schwalbe swallow
die Schweinerei (coll) dirty trick
schwenken (mil) to wheel
das Schwimmgeräusch, -e sound of swimming
die Seele soul
die Sehnsucht, -¨e longing
das Seil rope
sekkieren (S. Ger) to torment
der Selbstbetrug self-deception
das Selbstbewußtsein self-confidence
senkrecht vertical
die Sicherungsschraube, -n safety bolt
die Sitte, -n traditions
sittlich moral
sitzen, a, e (etw auf sich - lassen) (coll) to take sth lying down
sinnlich sensual
skizzieren to sketch
spannen (S. Ger) to notice, tumble
die Spannweite span, spectrum
spaßen to trifle, joke
die Spende donation
die Spendierhosen (pl) **(die - anhaben)** (coll) to be in a generous mood
der Spielverderber spoilsport
der Spießer petit bourgeois
spinnen (coll) to be mad
sich spitzen (auf jdn) (coll) to be keen on sb

spöttisch mocking
der **Sprachfundus** linguistic resources
sprachgewandt eloquent
das **Sprachniveau** linguistic competence
die **Sprachnot** inarticulateness
sprengen to chase
springen, a, u (coll) to work hard, jump to it
der **Sprung (auf einen -)** for a moment
spüren to feel, notice
ständig permanent, fixed
stauchen (coll) to give sb a dressing down
stechen, a, o to sting
stecken to hide
der **Steg** walkway
stehen, a, a (darauf -) to be the punishment
die **Stelle (auf der -)** on the spot
stellen to corner, catch
Stellung nehmen (zu etw) to comment on sth
das **Sterbekleid** shroud
stets always
der **Stich (im - lassen)** to leave in the lurch
der **Stier** bull
stoßen, ie, o to push
strafbar punishable
das **Strafexerzieren** (mil) punishment drill
stramm.stehen, a, a (mil) to stand to attention
strapaziert overworked, worn out
der **Straßenrand** curb, roadside
das **Streben** striving, greed
streiken to go on strike
sich **streiten, i, i** to argue
die **Strenge** strictness
der **Strich (gegen den -)** against the grain
die **Stütze** prop
der **Systemzwang, -¨e** pressure imposed by the system

die **Tagesschicht, -en** day shift
tastend tentative
tauchen to dive
täuschen to be deceptive
der **Themenbereich, -e** group of themes
Tonfärbung und Tonfall colour and tone of speech
die **Tonne** barrel, drum
der **Trab (in - setzen)** (coll) to get sb going
sich **trauen** to dare
treiben, ie, ie (es - mit jdm) (sl) to have it off
sich **trennen** to part
treusorgend devoted
triezen (coll) to torment, plague
das **Trinkgeld** tip
der **Tritt, -e** step
trostlos hopeless
der **Trumpf** trumps, superior

vom **Übel sein** to be a bad thing
überdauern to outlast
überdies moreover
Überfall! I'm being attacked!
überfallen, ie, a to assault
die **übergabe** opening
überhaupt anyway
überlegen superior
überschaubar small
die **Übersicht** view
überständig past it
überstehen, a, a to be over the worst
Überstunden (pl) overtime
übertrieben exaggerated, too much
übertünchen to cover up, whitewash
sich **überzeugen** to make sure
die **Überzeugung** conviction
über.ziehen, o, o (das Bett -) to change the bed linen
üblich customary
das **Ufer** river bank
um.bringen, a, a to kill
der **Umgang** contact
die **Umgangssprache** colloquial language
um.gehen, i, a to cope with

die Umkleidekabine, -n changing room
um.legen to kill
umreißen, i, i to outline
umsonst free of charge
der Umstand, -¨e circumstance
umstellt past help, lost, confused
um.tun, a, a (coll) to waste time
die Unabhängigkeit independence
unerlöst lost, without hope
unerträglich intolerable
die Unfähigkeit inability
ungehörig impertinent
die Ungerechtigkeit injustice
unheimlich eerie, sinister
das Unrecht (sich ins - setzen) to put oneself in the wrong
unselbständig dependent
unsichtbar invisible
unterbleiben, ie, ie not to be permitted
unterbrochen broken, interrupted
unterdrücken to oppress
die Unterhaltung entertainment
der Unteroffizier (mil) non-commissioned officer
der Unterrock petticoat
sich unterscheiden, ie, ie to differ from
sich unterstehen, a, a to dare
die Unterstellung, -en allegations
unverschämt rude; damned
unwillkürlich instinctive, involuntary
unzugänglich inaccessible
die Ursache cause
die Ursprünglichkeit originality
urwüchsig earthy

sich verabreden to arrange to meet
(sich) verändern to change
veranschaulichen to illustrate
der Verbeamtete, -n civil servant
sich verbeugen to bow
verbohrt pig-headed
das Verbrechen crime
der Verbrecher criminal
verbrecherisch criminal
die Verbrüderung drinking to friendship
der Verdacht suspicion
verdächtig suspicious
verdächtigen to suspect
verdammt! (coll) for God's sake!
die Verdichtungskraft power of compression
der Verdruß frustration
der Verehrer admirer
verewigen to immortalise
die Verfasserin authoress
das Verfehlen breakdown in communication
die Verflossene ex-girlfriend
zur Verfügung stehen to be at sb's disposal
sich vergeben (etw) not to do oneself any good, let oneself down
der Vergleichstext, -e text for comparison
das Vergnügen pleasure
das Verhalten behaviour
sich verhalten, ie, a to behave
das Verhältnis relationship
verheimlichen to keep secret
verhören to question
verkehrt wrong
die Verkümmerung, -en withering, atrophy
sich verlaufen, ie, au to disperse
verlegen to move; to set
die Verletzung, -en injury
sich verlieben to fall in love
der Verlust loss
vermeiden, ie, ie to avoid
vermeintlich alleged, supposed
verpassen to miss one's chance
verraten, ie, a to betray, let down
verratzen (coll) to be a goner
sich verrollen to roll, crawl into
die Versammlung, -en meeting, assembly
versaufen, o, o (coll) to drown
versäumen to miss
verschanzt barricaded
verschlucken to swallow up
verschlungen winding
die Verschraubung bolts, screws
verschwommen vague, hazy

versetzen (jdn) (coll) to stand sb up
die **Versklavung, -en** enslavement
die **Verspottung** mocking
verspüren to feel, have
der **Verstand (bei - sein)** to be sane
die **Verstörung, -en** distress
verstoßen rejected
verstummen to fall silent
verteidigen to defend
der **Verteidigungsgraben** defensive trench
vertiefen to deepen
vertragen, u, a to move
vertrauensselig trusting
vertraut familiar
vertreiben, ie, ie to drive out
vertun, a, a to waste
verübeln (jdm etw -) to take amiss
sich **verwandeln** to change
verwechseln (jdn) to mistake sb for sb else
verwundern to be surprising
sich **verziehen, o, o** to go away
der **Verzug** lateness, delay
vielmehr rather
die **Vielseitigkeit** variedness
der **Vogelpfiff** warning whistle
die **Volkspest** plague on the people
das **Volksstück** folk play
volkstümlich popular, colloquial
das **Volkswesen** spirit of the people
die **Vollständigkeit** completeness
von jeher all along, always
von mir aus as far as I'm concerned
die **Voraussetzung, -en** premiss
der **Vorbehalt, -e** reservation
vorbildlich exemplary
die **Vorgabe** disadvantage, handicap
vor.gehen, i, a to proceed
der **Vorgesetzte, -n** superior (officer)
vorhanden present
der **Vorhang, -¨e** curtain
das **Vorherrschen** prevalence
der **Vorläufer, -** forerunner
vorlaut impertinent
sich **vor.machen** (coll) to fool oneself
vornehmlich principally
das **Vorrecht** privilege
vor.sagen (jdm etw) (coll) to kid, tell stories
die **Vorschrift, -en** regulation
vor.schützen to feign
sich **vor.sehen, a, e** to watch out
sich **vor.stellen** to imagine
der **Vorstoß** attempt, application
vor.täuschen to fake, feign
vor.treiben, ie, ie to push forward
der **Vortrupp** (mil) advance party
vor.werfen, a, o (jdm etw) to blame, reproach
vor.ziehen, o, o to prefer

sich **wagen (in)** to venture into
wahr.nehmen, a, o to perceive, see
der **Wall** (mil) rampart
der **Weg (aus dem - gehen)** to avoid
weg.putzen (coll) to get rid of
sich **weg.schleichen, i, i** to slip away
weg.schnappen (coll) to steal
weg.tun, a, a (jdm etw) (coll) to insult, run sb down
sich **wehren** to defend oneself
die **Wehrmacht** army
die **Wehrpflicht** military service
weiblich feminine
die **Weiblichkeit** femininity
das **Weibsstück** (coll) cow, bitch
weis.machen (jdm etw) (coll) to make sb believe sth
der **Wellengang** swell, current
weltfern unworldly
das **Werbegeräusch** sounds to advertise product/service
das **Werkzeug** tools
wertvoll estimable, worthy
wesentlich significant
widerlegen to refute
der **Widerstand** resistance
wiehern to roar with laughter
der **Wind (in den - sprechen)** (coll) to waste one's breath
der **Wirbel, -** whirlpool
die **Wirkung** effect; influence
wirkungslos ineffective

wirr confused
der **Wirtschaftsfaktor** economic factor
der **Wirtschaftssegen** economic prosperity
wittern to scent
der **Wohnsitz** address, abode
die **Wortwahl** choice of word
der **Wüstling** lecher
die **Wut** anger

zäh tough, strong
die **Zeichnung** mark
zeitgenössisch contemporary
zerschunden tattered
der **Zeuge** witness
die **Zielsetzung, -en** aim
der **Zimmerherr** lodger
der **Zorn** wrath
zotisch smutty
zu.blinzeln (jdm) to wink at
zu.decken to cover
der **Zufall** chance
der **Zug (im - sein)** (coll) to have found one's rhythm, to have got something worked out
(in einem -) without a break
zu.geben, a, e to admit
zu.gehen, i, a to go on, proceed
das **Zugeständnis** concession
zugunsten (+gen) in favour of
zulässig acceptable
zu.lernen (- müssen) to have much to learn
die **Zumutung** demand
das **Zündholz** match
zurück.führen (auf) to trace back to
die **Zurückgebliebenheit** backwardness
zurück.stehen, a, a (hinter) to rank behind
das **Zurücktreten** reduced importance
zurück.weichen, i, i to shrink back
zusammen.basteln to cobble together
zusammen.fassen to summarise
sich **zusammen.nehmen, a, o** to pull oneself together
zusammen.richten (jdn) (coll) to straighten sb out
zusammen.stampfen (coll) to sort sb out, give sb what for
sich **zusammen.tun, a, a** to join together
zuschanden.stoßen, ie, o to wreck, ruin
der **Zuschauer, -** spectator
der **Zuschneider** cutter
sich **zu.schreiben, ie, ie (etw)** to be one's own fault
der **Zustand, -¨e** condition, state
zu.gestehen, a, a (jdm etw) to grant sb th
zu.treffen, a, o to be true
Zutritt verboten no entry
zu.wachsen, u, a to grow together
zuweilen at times
das **Zuwiderhandeln** violation
der **Zwang (unter -)** under duress
die **Zwickmühle** dilemma
zwingen, a, u to force
zwanglos relaxed
zwicken to pinch

THE SECRETARY
JOINT MATRICULATION BOARD
MANCHESTER
M15 6EU